Y0-DDR-037

SHADOWS OF AN AFTERNOON
A WAR AND ITS AFTERMATH

Shadows
of an
Afternoon

— War and its Aftermath —

M. L. GRAHAM

LUMINARE PRESS

WWW.LUMINAREPRESS.COM

Shadows of an Afternoon: War and its Aftermath
Copyright © 2024 by M. L. Graham

All rights reserved. This book or any portion thereof may not be reproduced or used
in any manner whatsoever without the express written permission of the publisher,
except for the use of brief quotations in a book review.

Printed in the United States of America

Luminare Press
442 Charnelton St.
Eugene, OR 97401
www.luminarepress.com

LCCN: 2024906491
ISBN: 979-8-88679-534-9

For my father, LeRoy G. Graham,
a member of the 1ˢᵗ Armored Division.

AFTER ANZIO

Rome

7 June 1944

Her passion pleasured Corporal Darryl Asquith even as it flew in the face of his military better judgment. Army regulations were being violated. Yet, to date, the war had been full of wrong-headed strategy and agonizing postponement, climaxing in a new iteration of *Dante's Inferno*. Violations or not, primal impulses had won the day. The soldier's coveralls were pushed to the ankles and his M-1 carbine rested in a hasty tilt against the bedroom's night-stand. He had not been with a woman in two and-a-half years, and the intimacy lacked coordinated cadence. But her tight hold on the American GI suggested she possessed a new-found freedom in finding love after the oppressive incursions of the German Wehrmacht.

The act had been spontaneous. The girl and her townspeople in this shabby, broken village on the outskirts of Rome had been subjected to war's fateful twists. Now she, and they, gleefully, welcomed Asquith and his 1ˢᵗ Armored Division—*Old Ironsides*—as it rumbled into the city in June of 1944. The city had been cleared—the Nazis chased. Though much of the town sat sideways in wreckage and townspeople were disgusted how the retreating Germans left the place, early June was upon them. There might still be time to plant gardens and hunt for a new future—indeterminate as it might be. At least there was now hope for a cleansed future.

Scents of garlic permeated the walls of her small flat, and a lazy ceiling fan struggled to move the stubborn, sultry air. Atop the nightstand a bottle of Sangiovese wine stood watch over the rifle—the weapon locked and loaded with more war to come in Italy; the jug, perhaps, fortifying both relief and courage for the girl to pick up the pieces, repair life, and move on after the fraudulent hegemony of Benito Mussolini. Caterina was her name—a thick language barrier made less cumbersome by the warmth of her smile. She had presented Corporal Asquith with a nosegay of flowers on the street below her casa. She looked to be in her late twenties, but war tested such calculation. She knew but a few words in English, those that might help her survive—perhaps knowing the same words in German when the Wehrmacht had occupied the area. The corporal had stopped to linger, not wishing to move jackal-like with his mates who were sniffing out vino and song in the nearby piazza. She had been standing beside a mustard-hued building pockmarked with bullet holes, and Asquith offered the best return he could upon accepting the girl's gift: a smile. He smelled the flowers and their eyes latched onto one another as soulmates might. Serendipity urged the two up to her dwelling where she offered him wine and hard cheese. Dialogue was mostly fumbled hand gesturing and more smiles, all leading to an easy embrace. Deeper intimacy followed.

Finished now, Caterina kissed his face then rose. She produced a moist, fragrant cloth and carefully washed him – she then tended herself. Asquith's natural reflex was to locate his rifle. For a short time, he had quit being a soldier—had in fact dropped his guard—the first time since *Operation Torch* cast him upon distant shores in 1942. Luck was running with him today as it had in Oran and Djebel El Hamra. The long and frustrating wait in the Padiglione Woods near Anzio beachhead had finally paid off for the Allies; Rome was liberated. But the division would have little time to drink in the celebration, as Asquith's unit would be pulling out soon.

Caterina dressed. Asquith hauled up his twill coveralls. They rendezvoused at the kitchen table and went back to work on the wine and cheese.

M. L. Graham

"Caterina…molto bella. My Italian, it's not so good," Asquith said. "Grazioso…you." He pointed to her.

Caterina's returned smile was self-conscious, shadowy, revealing a missing tooth in the deeper reaches of her mouth.

"You have family…ah, famiglia?" The soldier pointed to a photograph on the wall of an older man and woman. "Famiglia?"

"Si. Ah, you say, Morte, kaput…the Wehrmacht." She gestured a spit towards the floor.

"Bastards," Asquith said. He shook his head in disgust. "We, Americanos…we here now." He pointed to his rifle. "You be fine…"

Caterina's face stiffened, her smile a pensive one. She lowered her head and looked away.

For Asquith, life expectancy in this war had been thin as piss on a rock. Few rewards were to be had in battle—only luck and toleration. After basic training in Ft. Knox, Kentucky it was off to Ft. Dix, New Jersey for training on the half-track, insignia M-3: a 9-ton vehicle with traditional rubber tire front wheels and rear tank tracks. It was durable, weaponized, and designed to transport soldiers in an armored advance. The vehicle, known as the *purple heart box,* was constructed of quarter inch steel armor plate with stout front wheel suspension and volute springs making for a stabilized ride. It carried a driver and assistant driver with capability of hauling ten fully equipped soldiers in back. Firepower included a top turret 50 caliber Browning machine gun and two smaller caliber guns on side mounts. Although it had a utilitarian design, it also had two lethal setbacks: it lacked overhead air burst protection and was useless against machine gunfire directed at the unit's rear.

The corporal had been taught to shoot a rifle and schooled in the mental art of killing. After arriving in theater, he had done both and would never again be the same man he was upon leaving America. He, at least, had a weapon, and that tendered greater odds of survival, unlike this small village he now found himself in. Their only fortune was to have dodged the full rain of Allied fire from the previous week's artillery blitz as units pushed forward from

Anzio/Nettuno beaches and the Padiglione Woods in hot pursuit of the Wehrmacht.

The main component of the corporal's division arrived in North Africa in November 1942. It joined British expeditionary forces in Tunisia to push Erwin Rommel—known as the *Desert Fox*—out of Africa. The chase, in dust, mud, then more dust, had been a pursuit within the military's inherent "hurry and wait" slog—a slog harassed further by a defensive blitzkrieg of German 88s, one of the most dreaded artillery pieces of the war. Diving Stuka aircraft also figured in the mix, raining fire and death from above. Coming of age in Tunisia and its surrounds, Asquith quickly learned "military logic" to be fatuous, an existential contradiction of terms when applied to the world. *The world?* That was but a storied euphemism of life before this conflict. And now, Asquith found himself captured in a twisted juxtaposition of fate: having just made love to an Italian girl and his equal in knowing the horrors of war. Asquith's grungy, well used rifle revealed much of his story; Caterina's beautiful yet distant eyes, hers.

Fumbling interpretive exchange established the girl had no one left in Italy other than an older sister. Her three brothers, mother, father and various uncles and aunts had been rounded up by the Germans and "taken away." Asquith figured the only reason she had been spared were her age and attractiveness—her body and soul to be used as entertainment along the Wehrmacht's sham journey to glory. She seemed struggling to hold back weeping in describing what had befallen her, and Asquith knew her emotions were genuine and not wholly influenced by the strong wine. The war had hardened her—as it had him—and now, this frozen moment in time, this bizarre experiment in chance, had been consummated by spontaneous intimacy. Corporal Asquith wanted to hold onto the moment as long as he could but realized his unit had only a short time in Rome before resuming the chase of the enemy northward and into an uncertain summer.

The war was at a turning point. The day before (6 June 1944) the Allies had landed in France at Normandy. But the Nazi resolve would stiffen the closer they were pushed to Berlin, and the fight would become even more fierce—if that was possible. But for now, the soldier did not want to let go of Caterina and would spend the night—if she'd have him. Once again, the two embraced, meeting each other across the table.

Corporal Asquith emptied his pockets of Lira, the money issued by his central command to be spent in Rome in a noble gesture to keep the liberated city economically solvent. He gently folded some bills in Caterina's hand and pointed to the brick of cheese and bottle of wine, emphasizing in his stumbling Italian it was not for prostitution. Within the short hour he'd known her, he felt she surely understood that. She placed the Lira in her bag and, as Asquith reached for his rifle, she took his free arm and led him out of the casa and down into the piazza.

There the corporal encountered a couple of his mates, by now wobbly drunk. He ignored their bawdy inferences. The piazza had been overtaken by a festive air. Wine and song flowed. Dirty-faced kids ran roughshod, testing their begging skills with a pent-up banter most likely lost on former German occupiers. The two brand new lovers moved on, stopping at a farmer's market where Caterina expertly rummaged through a paucity of produce. There was no meat in the market; the Germans had seen to that in their retreat, taking away most stores and killing the remaining beasts indiscriminately. Asquith produced a couple cans of C-ration tins he found in one of his jumper's pockets and pointed playfully to his mouth. Caterina nodded approval.

The population had come out of hiding peacefully, thankfully believing the snowstorm of paper notes air-dropped 3 June by Allied aircraft urging them not to attempt defending the city for the Nazis. Some town's folk produced precious goods likely hoarded from the Germans and now for sale to inebriated GIs. Caterina continued her rummaging, selecting a leek, a parsnip, carrots and potatoes. She carefully doled out ten Lira, one unit of the denomination equaling one cent American. The young woman's bearings seemed reset now that some semblance of gaiety had returned to her Rome. Watching her in the simple tasks of shopping reintroduced sweet innocence to the hardened soldier. Something deep inside this quiet, unostentatious man was stirring; and it provoked even deeper feelings than those he had known with Helen—even in the last few days stateside when she had given herself to him. But Helen was a non-factor now. Her letters had become less frequent then stopped all together. It happened to many of the corporal's mates—a constant refrain related to war's bullets and bombs and long time away.

After the market, the two walked the promenade beside a small drainage canal. The stream emptied into the Tiber River, a main tributary flowing through Rome and from there to the Tyrrhenian Sea. The hubbub of the piazza behind them now, Caterina gently led Asquith to an unassuming cemetery where tree branches littered the grounds, most likely the result of the previous week's Allied shelling. They took a seat on a wrought iron bench in the shade of the afternoon, and the corporal lit a cigarette. She said many of her older relatives had been interred here. She crossed herself. Asquith, a Catholic, returned the gesture. The two sat in silence, her hand interlocked with his. Another soft embrace was followed by an exchange of deeper kissing. Asquith tasted the bitter tears that trickled down Caterina's face.

"You will be safe now, we Americanos here now," Asquith said.

"Ah, you soon go…I will be solitaria and no good."

"But we have the moment…ah, momento." He pointed to her then to himself.

"Si. Momento. But you soon partire…and I am sola and, as you say, very sad."

"It's the war." Asquith put his arm about her shoulder. "You are safe now…safe now."

"And you go to casa in Americano, and I am now here triste."

"I am sad too, Caterina. But we have today."

"And, we have stasera…we have tonight."

When the couple began to leave, a bird landed on one of the family grave markers. The bird, most likely a Eurasian nutcracker, had a peculiar rattling call. The bird seemed to be relating its own story, perhaps thankful quiet from all the shelling had once again returned to its habitat. On their way back to the girl's flat, the couple happened upon a bakery that had just produced a batch of freshly baked Italian bread. More Lira was exchanged. Next to the bakery a prominent village winemaker, Luigi Bianchi, jingled out of his basement cellar with a few bottles of vintage vino Caterina suggested he had hidden from the Germans. She and Luigi exchanged hugs. The man shook hands with the corporal, offering in quite good English he was so thankful for the liberation of the city. He offered the new lovers a bottle, gratis. Caterina thanked him and carefully placed the wine in her now bulging market sack. The two continued on, Caterina, happily toting her sack and an arm looped with Asquith's right arm—the carbine strapped to the corporal's other shoulder.

LATER AT THE CASA, ASQUITH EMERGED NAKED FROM THE GIRL'S copper bathtub, toweling himself. It was the first bath he'd had since the short R and R (Rest and Recuperation) in Rabat, Northwest Africa the previous year. Caterina was at her stove, the room's aroma reminding the soldier of home. A formidable stew using the Corporal's C-ration tins of ham and gravy and the market vegetables was in the making. The new bottle of wine had been opened and a jigger added to the simmering pot. Bread had been sliced and was on the table, a table dressed in a festive red cloth.

As the stew slowly warmed, Caterina approached Asquith and softly kissed him. She then led him back to her bedroom where she disrobed. War had stolen her innocence, but it could not take away her beauty. War had also postponed the young corporal's ideals of what love might be. The world conflict had delayed many important things in life—had, indeed, stolen life from many of his fellow soldiers. And, to begin to know love, Asquith first had to fight love's opposite—evil. He had done his duty to the best of his ability, refusing to think himself some kind of hero; rather, it had been a simple call to arms. Although he'd not been a complainer in the ranks, neither had he been an apologist. He did what he must, and this obligation to perform had made him deal with the devil at times. But now, he was over the moon with joy. It had been only hours since he met Caterina, but he knew Helen was but a ghostly image of the past, no longer relevant. What the young corporal was experiencing was no cheap carnal lust but true love. He sensed Caterina felt the same way. The two lovers once again found the bed and rolled passionately with one another.

Candlelight and lover's discourse, soft and hopeful, made this meal the best of Asquith's life. The war's reality was at a turning point. He had heard rumors. Many of the division's original boys were receiving orders to ship stateside and were rapidly being replaced by raw recruits. Might Asquith soon be one of the lucky ones?

"You come to America," Asquith said. "You will love America."

The girl slowly shook her head. "I, ah, I am Italiano, Si…but Americano, it simpatica."

"You will love it, Caterina. I, I from New York. I come from the pretty, ah, bella, the north…ah nord part of New York. It is green… ah, you say verde. I will come back for you."

Caterina smiled but looked down, some hesitation revealed in her dark brown eyes.

The dinner and wine lingered until dark fell upon the casa. The two then moved to the bedroom, made love once more and fell asleep.

THE ROOM WAS DARK WHEN THE CORPORAL CLIMBED FROM Caterina's bed and quietly dressed. He had composed a message the night before, found it in his jumper pocket, and placed it on the table. The note read: *I must return to my unit. I have spent the happiest day of my life with you. The war will soon be over, and I will come back for you. Please wait for me. Darryl*

Corporal Asquith put the photograph Caterina had given him into his pocket and collected the carbine. The soldier bent over and kissed Caterina on the cheek. She, still mostly asleep, twisted in an easy stretch, leaving Asquith that lasting image. He slipped out the door.

The corporal's 11th Armored Infantry Battalion, one of multiple components of 1st Armored Division, was mustering at its command post a short distance away. By 1600 hours he and his mates were once again aboard the *purple heart box* on their way up the leg of Italy. Rome behind him now, Corporal Asquith recounted his past year in—as the time-tested cliché went—**This Man's Army**.

RABAT, NORTHWEST AFRICA

7 June 1943

A swirl of humid wind kicked sand along a beach near Rabat, Morocco. Corporal Darryl Asquith's GI-issued swimming trunks, crafted of coarse-blend wool, were itchy and did not fit well. For the corporal, the swimsuit reinforced a *Sad Sack* absurdity only the military was capable of. Even in the relative calm of his R and R, the army's quartermaster found ways to screw things up. In spite of the suit's discomfort, the setting provided a break from the cold of the Tunisian theater and heat of battle that had taken place there. Just back from a plunge in the cool Atlantic Ocean, he dried himself with an olive-green towel and tried to forget the near disaster his 11th Armored Infantry Battalion experienced there.

The push for Tunisia in 1942 had been replete with leadership snafus and botched strategy by both combatants, especially at Kasserine Pass. In fact, that battle resulted in more an abandonment of the pass by the German Afrika Corp than a convincing Allied hammer blow. The Third Reich's brilliant German strategist, Erwin Rommel, known as the *Desert Fox,* had been outfoxed by someone on his own side: an impulsive Adolph Hitler. The Fuhrer had abruptly summoned the Field Commander back to the Wolf's Schanze deep in the Gierloz Forest of Poland right in the middle of the African campaign. This lapse in strategic thinking would be a hallmark of future Hitler tactical interference leading to the Fuhrer's eventual downfall. Rommel's recall provided the opportunity for the Allies to win the day by capturing 40,000 prisoners and inflicting a heavy defeat of the enemy in a most important

M. L. Graham

salient. Still, lots of mumbling within *Old Ironsides* felt they had not seen the last of the *Desert Fox*. At least the action at Kasserine Pass gave the now battle-tested 1st Armored confidence, something heretofore missing.

A change in command had also occurred on the Allied side. One of the first acts the new commander effected was a pushback of the unit to Rabat. More training and some much-needed R and R were in the works before the powers of war decided the division's next assignment. A large stock of the equipment had been left behind in Oran and Bizerte. It would provide important training stock and the opportunity of the newly consigned French to hone their skills. It had been a hard year's grind, and the men savored the "letting down of their hair" as the saying went.

First Armored set up bivouac in the cork forest south of Rabat. Men had been issued summer uniforms along with the swimming trunks to take advantage of the tropical Moroccan weather. The new uniforms were much needed. The old, insulated legging-type coveralls utilized by men in the armored branch for cold weather ops, were in shambles. The clothing had grown dangerously combustible as soldiers used pant legs to wipe everything from gear lube to C-ration ham and gravy grease on them. Corporal Asquith often used razor blades to scrape away the grease buildup then used it as an ignition starter for warming fires.

The beach at Rabat was a high point in the lows of war the rough and ragged troops of *Operation Torch* had encountered. Showers, hot chow, and even a few cases of beer were most welcomed. Some miscreants pressed further, letting their hair down way too far. VI Corp, a major player in the hierarchy of U.S. Fifth Army, provided MPs for monitoring the behavior of GI interlopers who might try trespassing the forbidden fortress of the Medina: the old African quarter where some unwitting men figured willing and able females might be stockpiled. Never could one short shrift the creativity of the American soldier when it came to figuring how to breach the Medina's walls. *Burnooses*—the hooded cloak worn by Arabs—were bartered with locals for C- rations and prized American cigarettes. The grey hooded cloaks were worn over uniforms allowing a few men at a time to pass through the Medina's gates. The *burnooses* would then be removed and tossed over the wall for the next group to enter. The charade worked for a time. But after a platoon or two of GI's curiosity had been satisfied—much to their frustration as Arab men were unwilling to share their women—the soldiers were apprehended trying to exit the Medina. Rounded up and tossed in the brig, the men were also strapped with a hefty $50 fine for their effort.

Corporal Asquith had no interest in seeing inside the Medina. He was a practicing Catholic and had left the proverbial girl behind. But war tested such orthodoxy. The previous six-months dreaded German 88 firepower had all but blown to smithereens any softening reminiscence of Schenectady, New York—and Helen. Now, the quiescent setting and easy routine snapped back happier images. Oh, how he remembered Helen's warm embraces, her sensuous whispers that night in the small hotel room he had hastily arranged before boarding the troop train in January 1942. In March 1942 it was on to Ft. Dix, New Jersey where he began a busy two-month training course in all things armored and 50-caliber weaponry. Dreams of Helen slowly began to fade. They ceased altogether aboard the *Queen Mary* in May 1942 as it steamed the 1st Armored Division to Ireland. In Ireland the men received advanced training for eventual staging to North Africa. He (and many other boys) had taken up smoking cigarettes. *Lucky Strike Greens* were standard GI issue, neatly packed in C-ration kits. Here he also developed a taste for strong, warm Irish beer. Alcohol's propensity to divert insecure thoughts of the unknown lasted until the hangover brought men back to reality. The division finished its training and staging in England then shipped to Oran in November 1942.

The corporal had quickly shed guilt for any bad habits of the soul contracted in Great Britain. War accorded such deals with the devil. But he could not shed guilt from an incident occurring at Djebel el Hamra in the Atlas Mountains of Tunisia in February 1943. There, he and PFC Guy Russell had been awarded a Bronze Star for valor. In Asquith's mind the whole thing had been nothing more than a spontaneous reaction, something that came natural in war. The pair had rescued three men trapped in a burning M-5 Stuart tank after it took a shell from German artillery. For Asquith, the citation rang hollow, as it held little compensation for the fourth crewmember that they were unable to rescue. The man's horrific screams as he burned alive in the depths of the Stuart now haunted Asquith, manifesting in nightmares and dismissing any bucolic aura of his previous life in upstate New York. He and Russell had grown closer after the incident. But befriending others in war carried its own risk. Asquith's 11th Armored Infantry Battalion command had urged longer serving cadre to welcome replacement personnel warmly, as it made for a more cohesive unit, critical when the firefight chips were down. After Djebel el Hamra, Asquith treaded lightly on such arrangements.

PFC Guy Russell, known as *Whitey*, road shot-gun seat in the corporal's half-track. An outspoken farm boy from Omaha, Nebraska who spoke fluent German (his German grandparents lived with the family), Whitey had undergone his own war zone transformation. German artillery and diving Stukas cast an indelible sheen of doubt on his own life's expectancy; one had only to do the math in the battle for Tunisia to calculate that. And the war would not be getting any easier the closer the Germans were chased to their homeland. Whitey was of sturdy, farm-hand build with thick, sandy colored hair, believed in God, and would have fit nicely into Asquith's ideals of a pal if it all had been happening in *the world*. They liked similar things: baseball (Asquith a Yankee fan and Whitey the Cubs); both twenty-two years old and owners of a jalopy with souped-up engine; Asquith a Catholic and Whitey an Episcopalian (close enough); both more than a little frustrated their love interests had been left behind and now subject to temptation from all the 4-F boys holding down the American home front. They had known each other since Ft. Dix, shared a bunk on the boat, and began their bad habits of the soul together in Ireland.

"So, what did you do with the Bronze Star, Quith?" Whitey asked, drying himself with a beach towel.

"What the hell am I gonna do with that...wear the goddamned thing? I tossed it. What'd you do with yours?"

"I sent it home. Mother would want it...especially if I don't get home.

"Don't give me that horseshit. You'll get home. I will too. If we survived Djebel el Hamra, we'll survive anything."

"I heard rumors we're going back east. Italy."

"Well, I know we're not going to Sicily. Second Armored's already enroute."

"I'm thinking southern Italy." Whitey lit a cigarette.

"After Kasserine Pass, I can't see how it can get any worse," Asquith said. He rummaged in his rucksack, finding a pair of aviator sunglasses and donning them. He then hauled out his cigarettes and lit one. As he stowed the smokes back in the pack, he found a heavily rumpled photograph of Helen and held it up to the sunlight.

"One fuckin' incoming would take you all out," intervened a booming baritone voice heavy with Okie accent. The soldier, owner of a tub-belly and looking ridiculous in an undersized swimsuit, stood over the two men, blocking out the sun. He was accompanied by a man still outfitted in his armored jumper.

"Up yours," Whitey said. He took a deep drag on his cigarette, offering a middle finger gesture.

The four laughed.

"How's the water fellas?" the second guy said.

"What the hell is it to you?" Asquith said. "Why are you still overdressed? Get out of that damned jumper and go swimming like the rest of us."

The new voices belonged to Jimmy *Dago* Esposito and a kid named Sydney Beemer, or *Spud*—the nickname owing to his Idaho hometown. The two had also trained with Asquith at Ft. Dix and in Great Britain.

Esposito, a corker, was imbued with a natural arrogance that matched his jutting brow and combative jaw. His personality could easily incite a bar brawl (and had at Ft. Dix). With two stout fists, his points were easily taken. Hailing from St. Louis, Missouri, he was older than most in 11th Armored Infantry Battalion. Dago Jimmy had gotten his buck sergeant stripes just as the division moved to North Africa. Although his Okie accent cast the fellow as a backwater lout, he was someone you wanted to have around. Dago had quickly developed a knack for killing Krauts and possessed the clever ability to disarm landmines, a constant threat in armored advances.

Spud was a mild-mannered Mormon from Preston, Idaho who prayed a lot. As is typical asymmetry in *this man's army*, the pairing of Dago Jimmy and Spud was a showcase of polar opposites; and as is a proven postulate in physics, the two men attracted, bonding as a band of brothers. Spud, a PFC, resisted bad habits of the soul infecting most of the others. For this lad, God shadowed his journey along the ever-narrowing path of fate. He refrained from rough language, common among most of the Eleventh. He had no critique for the division's inconsistencies of command; did what he was told; and he carried no crumpled photo of the girl left behind. His most cherished possession was a well-worn *Book of Mormon*, the standard works of his belief.

This unlikely foursome sat on the beach shooting the shit in the warmth of an equatorial sun. When Dago Jimmy became annoyed with Whitey fingering his dog tags, he offered, "You sure play with those damned things a lot."

"Whadda you suppose this notch on the dog tags is for?" asked Whitey.

"Hell, if I know," Asquith said.

"I been using it to clean my fingernails," Whitey said.

"God you morons are dense," Dago Jimmy scoffed. "Christ, it's a tooth notch used as a pry to keep your mouth open so the tag doesn't get lost and arrives with the dead stiff when he's shipped to the rear for burial—that is if you're so lucky to have any face left."

"You gotta be shitin' me," Whitey said.

"I ain't shitin' ya. Spud and me had to help the corpsmen get Billingsley out of that bomb crater up at Sidi-bou-Zid. Ole Billingsley didn't have no legs left but at least still had some face." Dago Jimmy took his dog tag and wedged it into his mouth. "See, it works like this."

"Oh, shit, that's enough you hick," Whitey said.

Corporal Asquith looked at his dog tags and said, "For heck sake."

"Well, ain't this just fine chit-chat," Whitey said. "You're spoiling my R and R." He looked over at Spud who was lying on his side, propped up by an elbow. "Strip off, Spud. The water's fine."

"Leave the kid alone," Dago Jimmy said, awkwardly gaining his feet. "I'm heading to the water to get away from you lose screws."

Asquith extinguished his cigarette, mashing it in the sand. "You know just how funny you look in that swimsuit, Dago?" he said.

Dago Jimmy looked back and said, "Fuck you!"

Laugher hung in the close air.

WITH TIME SPENT IN THE RELATIVE CALM OF RABAT, MOROCCO in early summer 1943, Corporal Asquith had time to reflect upon the incident that had occurred near Djebel el Hamra the previous February. It had changed his life. He had seen the flash and felt the concussive impact simultaneously. The blast knocked him and Whitey to the ground where they had been standing beside their half-track. Natural reaction was to turn away from the noise and heat, but they could not dodge the stinging gust of dirt clods ricocheting off helmets and the sides of their faces. A small chunk of flesh dislodged from the back of the corporal's right hand, a hand that had been instinctively positioned to guard his face. Blood splattered across the sleeve of his field jacket and onto the sides of the half-track. Before either man could utter a word, another shell hit, and Asquith knew from the sound and impact it was a German 170 mm howitzer. That burst caused an echoey secondary explosion.

By the time the dust and smoke cleared, and the men regained some sense of time and place, they realized the M-5 Stuart tank positioned beside their half-track had taken a direct hit. Reddish yellow flames lapped from the tank's unraveled track runners, and the rotating top turret sat sideways and at a tilt.

Eleventh Armored Infantry Battalion, combined with Combat Command B, both units' part of 1st Armored Division, found itself encircled by the enemy in the Tunisia salient some thirty miles from the previous day's action at Haidra. The Eleventh had been given far too much territory to handle by itself, and the Germans had picked up their range. It was 1315 hours, 21 February 1943.

The battered hatch of the M-5 had been blown open and smoke billowed out. The explosion had catapulted one crewman from the turret, and he lay burning on the side of the tank. Instinctively, Asquith and Whitey raced to his aid, Asquith thinking to grab a shelter half in back of his half-track. The flames were quickly smothered by the canvas. Another two soldiers raced over and dragged the man away. Next, the corporal and PFC leapt upon the hull of the tank, ripping the hatch further open. Asquith could see the tank commander still in the depths of the Stuart. He was cooking and from the looks of his shredded overalls, had major wounds. Incredibly, he was still conscious and was wrestling with another crewman—the bow gunner. Asquith leaned through the hatch opening and was able to grab the gunner's field jacket near the nap of his neck. As the corporal pulled, the commander pushed, and the man was out of the tank. Whitey dragged the man off the tank and away. Still in the tank's depths, another man struggled to free himself. It was the assistant driver. The tank commander shouted that the man's foot was pinned in the wreckage. Asquith shimmied through the turret's opening and into the tank hold where he and the commander tried to free the driver. Choking smoke and flames worsened and the commander was quickly losing consciousness. Asquith put a full-Nelson on the man and used all his might to drag him to the turret opening. He then bearhugged the commander, pushing him up through the hatch where Whitey pulled him from the top. He was out.

"There's one more in here," Corporal Asquith yelled, taking in a deep breath of fresh air and dropping back inside the tank.

"Get the hell out of there, Quith!" Whitey shouted. "It's gonna blow!"

Smoke and heat had made the environment untenable. Once more Asquith tried to free the man from his inevitable tomb. Whitey again screamed for his corporal to get out of the tank. Finally, Asquith struggled out empty handed, coughing so violently he puked.

A couple other members of the Eleventh helped Asquith off the tank. As the trapped soldier screamed in agony, the men hurried away fearing an explosion from the tank's fuel and ammo loads. Out of the smoke and chaos another soldier emerged, rushing up in a low crouch. The man had two bars on his collar and carried a tommy gun. His name, Captain Paul Shields.

"Captain, there's still a guy in there," Asquith shouted.

"Can't he get out?" the captain yelled.

"His foot's stuck in the wreckage."

"It's gonna blow, Corporal...there's nothin' that can be done."

"We just can't let him fry in there," PFC Russel shouted.

"Are you sure we can't get him out?" The captain again asked.

"He's leg's stuck under twisted piping." Asquith said, continuing to gasp for air. "We got to get the fire out."

"There's no way we can do that Corporal," Shields yelled. Then, with a calculating tone, the captain said, "Well, I can do something." He sprinted towards the tank, leapt upon the hull, stuck his tommy gun through the hatch and fired a short burst. The screaming stopped. The incident at Djebel el Hamra had taken less than ten minutes to playout but would last a lifetime for Corporal Asquith.

M. L. Graham

ANZIO BITCHHEAD [SIC]

January 1944

"On your feet, men, we're movin' out," Captain Shields ordered as he made morning rounds, the omnipresent tommy gun slung about his shoulder.

The army at dawn broke its slumber, and Corporal Asquith's recurring nightmare had ended. The order to move out did not come as a surprise; rumors had been running rampant. An opus of moans, further orchestrated by choruses of mumbled expletives, sounded the length and breadth of 11th Armored Infantry's shelter-half encampment. R and R for *Old Ironsides* was at an end and another chapter in war's great unknown was about to begin. A sizable line of dog-faced soldiers, many dressed only in white under-briefs and combat boots, mustered at honey pots for first order of business. With three months in Rabat and decent hot chow came the bonus of predictable bowel movements. Chow was the next line to form. Clanging mess kits and smells of first-of-the-day Lucky Strikes competed with rattling cook spoons and the enticing whiffs of coffee and bacon. Absent screaming Stukas and titillated by aromas of *the world*, the division had been getting far too comfortable—that was dangerous.

By 0800 hours bivouac amongst the twist of ghoulish cork trees in Rabat was fully decamped. G-2 reported Sicily now secured, the Germans on the run north. The foot of Italy would be the next destination in *Operation Shingle* after a non-tactical journey back to Oran. From there the division would pile into LSTs (Landing Ships, Tanks) for Naples. Then, perhaps, 1st Armored Division would tease its way up the shapely leg of Italy; but that was anybody's guess. The only thing certain in *this man's army* was uncertainty. For the GI Joe's of *Old Ironsides*, fate—as it had been in Djebel el Hamra and Kasserine Pass—would determine if they'd ever again reunite with *the world*—body and soul intact.

Questionable strategy clouded the concept for *Operation Shingle*. The plan came to life at the Tehran Conference held at Casablanca in late November 1943. That conference, attended by U. S. President, Franklin Roosevelt, British Prime Minister, Winston Churchill, and Soviet Premier, Joseph Stalin, set forth a commitment of Western Allies to open a second front against Nazi Germany. Primary push would be Operation *Overlord*, the long-planned cross-channel invasion of Western France. Additional strategy included the more controversial idea to liberate Italy. The Italy plan had three main reasons: 1) Bind German forces to keep them away from the Eastern Front; 2) Provide airbases in which to bomb deeper into Germany; and 3) Make it easier to use seasoned troops to later participate in *Anvil*, the push into Southern France in coordination with *Overlord*.

With *Operation Shingle*'s final approval, the main body of the 1st Armored Division and 43rd Infantry Division would set up in Naples. D-day's planned landing on Anzio/Nettuno beachhead would come late in January. The operation would use LSTs to drop off men and cargo, then scurry the craft away to England in preparation for Normandy landings, then scheduled for May 1944. Allied bombardment of artillery, rockets, and naval gunfire preceded troop arrival in Anzio. Initial landing, 22-28 January 1944, caught the Germans off guard. Calm seas helped, but there was potential for treachery as the Germans had scuttled ships in the bay to forestall any invasions and pesky sandbars had to be factored in. Once ashore there were more concerns. A narrow plain opened to the east, buttressed in the distance by Colli Laziali and Lepini Mountains, rising nearly 3,000 feet in summit. This provided the Wehrmacht with excellent defense lines.

M. L. Graham

Men and equipment—approximately 90,000 troops and 100 tanks and various other armored apparatus—unloaded at Anzio/Nettuno beachhead. Included in the count was Corporal Asquith's M-3 half-track which moved ashore without any significant resistance. Eleventh Armored Infantry Battalion quickly located inland between the Moletta River and the Mussolini Canal—the canal was a public works project campaigned by the dictator for controlling the Tiber River delta in an effort to increase buildable land. For Asquith thus began a miserable five-month cat and mouse bivouac in the Padiglione Woods—a tangled underbrush and scrub forest about four miles inland from the beach. Winter rains made for a veritable slog—this paired with nettling harassment by both mosquitos and *Axis Sally's* radio broadcasts from Rome. It was as if she knew the exact coordinates of equipment bogged down in a quagmire. *Midge the Bitch* figured into the bothersome mix as well. The only relief of her radio broadcasts was the frequent playing of *Lili Marlene,* a German tune Asquith had grown fond of.

Once established ashore, there was no escaping the blitz of German artillery. Slit trenching and fox-hole-digging was arduous, owing to sandstone underlayment extending inland from the beach. With resulting shallowness of cover, the men got creative using wooden crates from artillery shells, C and K ration cardboard, and shelter-half layers laminated beneath camouflage netting to make their new home homey as possible.

The division's first action came 29-30 January. Objective of 1st Armored advance was to join a contingent of the British 179th and one squadron of 46th Royal Tanks in a push to effect a cut into Highway 7 at Cisterna—an important junction towards Campoleone; from there movement into the Colli Laziali hills might be possible. A second objective was an advance to Campoleone itself where an important rail hub allowed the enemy to resupply. No question the Allies would meet with stiff enemy resistance there.

In that effort, a leading column of armor pulled off the Albano-Anzio Road and moved almost a mile to the northwest sector.

There, three 46th Royal Churchill VII tanks and two British half-tracks bogged down. The enemy spotted the stalled units and opened fire. Corporal Asquith, who was directly behind, diverted his half-track, pulling his *purple heart box* into a dense thicket that offered good camouflage. As German 88s picked up range, he and Whitey and half-dozen infantry troops they were hauling bailed, racing in a low crouch to the protection of a roadside ditch. The Churchills were fortunate when the Allies returned fire as a battery of 105 mm howitzers on the beach found range, scattering enemy guns. The resulting fire power allowed the British to remove their bogged equipment, thus escaping with minor casualties. Asquith and his crew also took advantage, repopulating their half-track and—as the sarcastic aphorism in the division went after Kasserine Pass—advanced to the rear.

Next day, a concerted drive to Cisterna with reinforcements from the US 3rd Infantry Division went nowhere. An attached Ranger Force tried a night approach but ran square into an ambush. Only six of seven hundred men came back. The enemy had regrouped quickly and was much stronger than originally thought. Most Allied advances halted, and the division dug in.

The enemy's most fervent counterattack came on 3 and 4 February. The Germans air-bombed the beach, hitting the 95th Evacuation Hospital which housed a dozen 16 X 20 foot red-cross-marked tents. The evac facility was hard hit, suffering 23 killed. The Germans quickly assembled strong forces in a gully where Asquith's 11th Armored Infantry Battalion had made its first attack. Combat Command A moved up tanks behind the Campoleone-Cisterna railroad embankment for deployment down Carano Road, a strategic route factoring in Wehrmacht defense. In the night another Allied line was established north of German outposts, Carroceto and Aprilia (*The Factory*), so named by the shape of its buildings. The advance stalled. On 7-8 February the Germans resumed its offensive along the Albano Road. British 1st Armored defended the position, but its troops were fatigued, and the enemy hit them on both flanks; by 9-10 February the British were exhausted and in trouble, and Asquith's attached 11th Armored Infantry Battalion received heavy assault from the German 105s and a Panzer division of Tiger II tanks and motorized infantry. Again, the Allies dug in.

The Germans then launched its main effort to retake Anzio beach on 16 February. During an incident near Carroceto, a Cub spotter aircraft found a couple thousand enemy and provided coordinates. Allied artillery pounded the area, and one observer was to have said the Krauts had been knocked over like bowling pins. The place came to be known as the Bowling Alley. Bowling pins or no, Allied high command felt not enough was getting done fast enough. By 23 February a change in leadership for *Operation Shingle* had occurred. But good leadership had no say in the weather, which was winning the day. Stalemate ensued. First Armored orders were to withdraw the division, including Corporal Asquith's 11th Armored Infantry Battalion, and fallback to the Padiglione Woods. There the long wait at Anzio had begun.

INTERMINABLE WERE THE WEEKS LEADING UP TO THE PLANNED spring breakout. Miserably dank in their shallow foxholes, a shivering Asquith and Whitey perfected the art of keeping their heads down. Next hole over, and within earshot, were Dago Jimmy and Spud. At night troops listened to the paradoxically addictive *Axis Sally* and *Midge the Bitch* broadcasts from Rome.

"Hello suckers," *Midge* loved to taunt. "Are you missing your Frualein back home, Joe? Johnny Boy's with her now. Too bad, Suckers." The females' assaults over the airwaves were interrupted by whistles of incoming German 170 mm howitzers, climaxing in hefty explosions. It was hell of a way to kill time—and "killing time" offered soldiers greater opportunity to contemplate the possibility of the real thing when better weather would make possible a full-frontal assault to free Rome.

"Hey, Quith," Dago Jimmy called between shell bursts. "Heard you got a letter from your gal. I hear 4-F Johnny's sniffing around her now, huh? At least that's what *Midge the Bitch* just suggested."

Muffled laughter rose from surrounding foxholes. Corporal Asquith hard-balled a C-ration tin of chipped beef, and it ricocheted off Dago Jimmy's steel pot helmet with reverberating clunk. Another chorus of laughter lifted into the smoky air.

"Hey! Pipe down over there," ordered Captain Shields, two holes beyond. "And if you're gonna smoke for Chrissakes Jimmy, do it under your shelter half. Jerry will have our range when they see the cigarette's glow."

Dago Jimmy tucked his head beneath the shelter-half and continued to work his cigarette. Spud coughed, poking his nose out the other side of the foxhole for fresh air.

The corporal had, indeed, gotten a "Dear John" letter and had made the mistake of mentioning it to Whitey who must have spilled the beans to Dago Jimmy. This was not the land of tender feelings. War was raw, and men learned quickly to keep news of home, good and bad, to themselves. But it was the confounded waiting that generated tensions. A couple half-track jockeys from 11th Armored Infantry had duked it out with a pair of hotshot tank boys out of Combat Command A. Captain Shields had been nearby and broke up the fist fight. The encampment was on tenterhooks as the bastard Wehrmacht refused to come out in the open and fight.

Anzio Beachhead and the Padiglione Woods were most restless at night. Potshots from snipers periodically found their marks. Enemy planes dropped flares, and light filtered deep into the woods. From sundown to daybreak the Allies retaliated, going after both specific targets spotted in the day by Cub air observers and random lobs on suspected enemy encampments. Even though many men had been in the field since 1942, for most it was impossible to sleep—unless one had scored some Italian contraband brandy, gotten blitzed, and passed out cold. One infantry grunt from Dago Jimmy's bunch, a Corporal Gleason from Atlantic City, New Jersey, got so drunk he passed out in the woods and not only slept through a fusillade of enemy shelling, but the cigarette he had between his index and middle finger smoldered right through the skin to the bone. He had to report to 95th Med-Evac for treatment and was out of action nearly a week. Captain Shields found out he had been drunk and sent the man from corporal to buck private before the hangover had worn off.

Many men found God, bombs and bullets having a knack for swelling the ranks of first-time believers. Some agnostics figured they might as well take a chance on a Maker too—what did they have to lose? After all, they were flexible in philosophical bent, well-bred for compromise. Atheists, more rigid, spouted hard-worked excuses—an incorrigible lot, these boys. Chaplain Anderson of 16th Armored Engineers was a stalwart for many men, those in action, the wounded, and the dead. He went to great lengths to organize worship services as well as programs for recreation and sports. In the Padiglione Woods the resourcefulness of *Old Ironsides* shined in the construction of a "Church-in-the-Wildwood" built from trees felled by enemy fire. On Easter Sunday a dawn bugler sounded the clarion call for soldiers far and wide to come worship. Later that day, at the beachhead, another service for Catholic and Protestant worshipers was marred by incoming German 170 mm howitzers, scattering the congregation. That attack also scored a direct hit on *Little Cleveland*, a bivouac for an attachment of Ohio National Guardsmen. Ironically, attendees from *Little Cleveland* at the "Church-in-the-Wildwood" survived the shelling while many Ohio atheists who stayed in camp did not—perhaps proof-positive for the existence of God. War confounded metaphysics.

Still the wait continued. When local farmers were evacuated to Naples for their own safety, GI's took charge of their cows. Some bovine were used for milk and others for beef steak. Some enterprising Italian women were conscripted to wash GI's clothes— sergeants and above assigned as morality monitors in an effort to ensure wet washing did not lead to prostitution. A black market for brandy arose but had high risk—just ask Private Gleason. As in Rabat's Medina, GIs with too much time on their hands could not resist attempts at perfecting genius; but with little outlets to test their schemes, bullshitting proved the best outlet, surfacing most in crap games. Fortunes were won and lost here; however, as this was an army in wartime with limited personal finances, fortune was a relative term.

In March 1944 major replacement units began to arrive, continuing well into April and May. The 45th Infantry Division drew much needed personnel. The British 1st Armored Division hustled in a new brigade and the U S 34th Infantry Division came up from the south after a brief R and R. With the return of borrowed men, the 1st Armored Division had both Combat Command A and B back in the fold. Full-frontal skirmish, and all it portended, was locked and loaded.

BREAKOUT

May to June 1944

Early in May resupply problems were solved by obtaining additional landing craft and bringing in more than 14,000 replacement personnel. This registered nicely with improving weather. However, on 5 May a controversy arose disjointing high command. It seemed British Field Marshal Harold Alexander had persuaded Major General Truscott of 1st Armored to implement *Plan Buffalo:* a push of Allied forces toward the Campoleone-Cisterna-Terracina railroad salient in an effort to cut off enemy supply lines crossing the Velletri Gap. Fuming, 5th Army's overall commander, Lt. General Mark Clark, insinuated Alexander had interfered in his chain-of-command. It was no secret Clark wanted a more direct entry to Rome; and the sooner the better. Symbolism was important and would achieve an Allied psychological victory in *Operation Shingle;* and it would carry over nicely to the Normandy invasion of *Operation Overlord* as well. Clark thought *Plan Buffalo* might hinder that, and his thinking flew in the face of a contrary argument holding that a rush to Rome would fail to account for the strategic opportunities in cutting off the Wehrmacht as he escaped up the leg of Italy. As the third week in May approached, the decision on *Plan Buffalo* had still not been confirmed.

A beam of light flashed across Corporal Asquith's face, awakening him in a start. After a recent reconnaissance mission with Captain Shields, the two squad leaders, Asquith and Dago Jimmy Esposito, had relocated their men to the northeastern edge of the Padiglione Woods preparatory Allied breakout scheduled for this day, 21 May. The corporal, along with PFC Whitey Russell, were sharing a slit-trench with Sgt. Esposito and PFC Spud Beemer. The spark of light had come from Spud's flashlight, the kid reading his *Book of Mormon*. He seemed always to be reading it. Such obsessive habits were not rare occurrences with men in war facing the likelihood of meeting their Maker. Spud's fixation, however, succeeded in raising the ire of Whitey, who owned contrary religious views. If harassment by lethal, sharp-shooting snipers and artillery blitzes by the relentless Hun were not bad enough conflict, now a religious war was brewing.

"Easy with the light, Spud," Asquith said. "Jerry might be watching." The verbal offering awoke the others.

"What the hell's happening?" Dago Jimmy spluttered. He stuck his head out of his sleeping bag. "Shit, and I was right in the middle of a dream...and it was a good one if you get my drift."

"What was her name?" Asquith asked.

"Don't know...never saw her before. Tell ya, though, built like a brick shit house...stacked, whoa my god."

"Sounds like a Betty Grable pin-up," Asquith said.

"Who was shining that flashlight"? Dago Jimmy asked.

"Just Spud reading his book," Asquith said.

"Figured it was the Reverend," Whitey said. He rubbed his eyes. "Now everybody's awake. What the hell do you see in that heathen book? Try the *New Testament* sometime."

"I'd trade both those goddamned books for a pinup rag of *Yank Magazine* about now," Jimmy said. "Got me a hard-on the likes of the second comin.'"

"Get the hell out of my foxhole with that sorry thing," Whitey said.

Asquith could not help but think such foxhole banter borderline macabre, what with the great battle lying ahead.

"Sorry, men," Spud offered. He tucked the book away in his field jacket and put the flashlight into his rucksack.

"Is that book something like the *Holy Bible*, Spud?" Asquith asked.

"It is," Spud said with proselytizing excitement. "It's about persecution of people in the land of Israel and their long journey to new lands in the Americas."

"Persecution? Don't get me going on that topic," Dago Jimmy said. "Persecution is having a hard prick in the middle of a war zone and nowhere to stick it."

"Well, I know where you can stick it," Whitey shot.

"Yeah, well…"

"You know the only genuine book on the subject of Jesus is the Bible… the *New Testament*." Whitey said.

"Whitey, you really ought to give this book a try, It's, well, it's fascinating."

"I don't think so, son. That shit. I've see'd it sold by snake oil salesmen who come through Omaha once. Heard 'em Mormons have horns and bunches of wives. It's called polygamy…now, don't they?"

"What?" Asquith said. "They have more than one wife?"

"Sure do," Whitey said. "Shit, sort of like 'em sultans in the Medina back at Rabat."

"Man, what I'd give for just one wife about now in place of you lose screws," Dago Jimmy said.

"Oh no, the keeping of multiple wives is not practiced anymore. A revelation saw to that," Spud said.

"Revelation?" Asquith asked. "So, that's like talkin' to God?"

"Sure, why if…," Spud's reply was interrupted by Dago Jimmy.

"Good god! On the day we break out of these sorry woods to battle Jerry, got me a boner the likes of kingdom come, and these two morons talkin' about who's got the best religion book."

"Shut the fuck up over there," a voice rose from a neighboring fox hole.

"Up your ass too!" was Dago Jimmy's return fire.

"I'm sorry, guys," Spud said. "I'll quit bothering you about it."

"Ah, it's okay kid, it's okay," said a relenting Dago Jimmy. "We all need help gettin' through this war. And I gotta feelin' the shit's really gonna hit the fan with this breakout. Still, the sooner it's over, the sooner my dick's gonna be a happy stiff back stateside."

"You're a sorry son-of-a-bitch, you realize that don't cha Dago Jimmy," Whitey said.

"Well, I try, boys, I try." The curmudgeon from Missouri fired up a cigarette, filling the slit-trench with smoke.

Corporal Asquith parted the edge of the foxhole's shelter-half and searched the horizon. A tangle of light was becoming visible through thickets of dusky green conifer. Darkness was losing its grip with yesterday. And today, today the long wait in the Padiglione Woods would be at an end.

PLAN BUFFALO IT WAS. THE ARRANGEMENT WOULD MAKE AN advance in the direction of the Velletri Gap in an effort to snag the Wehrmacht's vulnerable communication links. It would be risky to create such a deep salient there—guarded on three sides by the enemy—but if successful, it might make way for a pivotal cut into Highway 7—the road to Rome—and well worth the effort. A week before, artillery from VI Corp provided daily doses of fire power, becoming more intense as D-Day approached. The main drive to the northeast would rely on 1st Armored and 3rd Infantry Divisions providing major assaults with 45th Infantry protecting the west flank; Special Forces would do the same on the southeastern side. So, on 21 May Major General Harmon issued Field Order # 10. The order would dispatch 1st Armored Combat Command A toward Cisterna and the Velletri Gap with one battalion of medium tanks, a battalion of light tanks, two battalions of infantry with one infantry battalion left in reserve.

After a day's delay, units moved out at 0630 hours on 22 May in grey, overcast, a day unable to be counted on for Allied air support. Caught off guard, a small gathering of enemy was rounded up or shot as they huddled inside ditches. Engineers marked and removed landmines as the advance continued. Third Battalion of the 1st Armored Regiment did a nice job using *Snakes* to clear the way for armored advance. (Snake*s* were long metal tubes loaded with high explosives invented by 16th Armored Engineers to clear the way by blowing up landmines). With aid of forward observers, supporting artillery fire was a tremendous help in Combat Command A's advance. By high-noon medium tanks were nearing their railroad objective. By mid-afternoon the attack had gone past the railroad and by the time dusk set in, Combat Command A had gone over 500 meters into enemy territory.

However, things had not fared well for 1st Armored's Combat Command B, which included Asquith's 11th Armored Infantry Battalion, as they pushed toward Campoleone. One reason was the decision not to use *Snakes* to clear their path ahead. In a short time, many tanks hit landmines and blew off their treads. They now sat immobilized. Telecommunication had not been executed properly, and the damaged equipment remained out of service awaiting recovery. Armored infantry units, which included Asquith and Dago Jimmy's half-tracks, pulled back to await tank recovery efforts. It was at this point the war took a fateful turn, casting a heavy pall upon all those who knew Sgt. Dago Jimmy Esposito.

Captain Shields had been ordered to assist sappers (minesweepers) with Company C of 16th Armored Engineers in tracking and neutralizing landmines; it would greatly facilitate the recovery effort of Command B's Shermans. The natural choice was Sgt. Esposito. The sergeant's half-track with Captain Shields riding shotgun ventured ahead, accompanied by Asquith's unit following a short distance behind. German artillery had taken a break, either in retreat or simply not seeing the sitting-duck tanks of Command B. Company C was already on scene. They had sent out sappers with mine wands sweeping a safe haven along the tanks' starboard flanks. This would insure a secure quadrant for the tank recovery mission.

The captain addressed the detachment. "Sgt. Esposito, you and Beemer and your squad help the sappers of Company C. They'll do the spotting. You guys do what they say. Any defusing of the landmines from us I want only Esposito doing it; he's got the training. The rest of your and Asquith's squad be runners for equipment and on the lookout for snipers." Captain Shields walked with Asquith's squad back to his half-track to get shovels. There was an explosion. "What the hell," the captain yelled. "What the hell!"

Thinking it Jerry's incoming, Asquith and Whitey dove for cover beside their half-track. As it turned out, the blast was a boo-by-trap landmine, planted at the edge of the roadway as subterfuge to nail unsuspecting sweepers. The blast caught Sgt. Esposito in the legs and chest, blowing away both feet and ripping open his rib cage. The big lug from Missouri died instantly. Spud Beemer, who had been a few steps behind his sergeant and somewhat shielded, was hit with a glancing blow of shrapnel. The blast tore into his field jacket, scoring a direct hit to the *Book of Mormon* carried in its pocket. The book was ripped to shreds, pages fluttering across the blast zone.

Asquith and Whitey recovered quickly, hurrying to the scene as a platoon leader from Company C ordered them to halt. The scene was not clear. Sappers moved in quickly to sweep the area. In a matter of moments, the site was declared safe.

"Get a medic up here!" shouted Asquith.

The explosion had quickly mustered a squad of medical corps-men. "We clear?" one of the medics hollered. "Affirmative," some-one yelled. The medics moved in.

Corporal Asquith reached the big Missourian. The blast had exposed his internal organs in a twist of gore. Whitey pushed in. The medic shook his head and extended his hand to halt. "He ain't gonna be saved, boys," the corpsman said.

"Jesus Christ, man!" Whitey screamed.

"Bastards…bastards," Asquith shouted.

Spud crawled closer to his sergeant. "Jimmy, Jimmy," he called. The devout Mormon from Preston, Idaho had suffered a gash on his forehead and a glancing wound on his right side. A corpsman swooped in to begin treatment.

M. L. Graham

"Get my book, Whitey," Spud called, "get my book." Spud's voice trailed off, the young soldier in obvious hypovolemic shock.

The medic cut away Spud's still smoldering field jacket. "Whatever you had in that pocket deflected much of the shrapnel," he said. "Probably saved your life troop."

Dazed, Asquith and Whitey gathered as many pages of the book as possible, most blood-stained. The corpsmen applied trauma dressing to Spud's wound as Asquith stuffed the book's pages into Spud's rucksack found in the back of the nearby half-track. Asquith then moved to where Captain Shields stooped beside Dago Jimmy.

Asquith regained his feet and found his M-1 carbine, donned his steel pot helmet and walked back to the half-track in silence.

"How 'bout that book saving the kid's life," Whitey said, joining the corporal. "The Lord sure works in mysterious ways."

"Bastard Krauts," the captain said, accepting a blanket from another corpsmen. He covered the lower half of the sergeant's body then untangled Esposito's dog tags from around his neck. He separated the tags from the attached chain and put one in the sergeant's field jacket coat and kept the other. That task complete, the captain pulled the blanket over his body.

"You didn't put one tag in his mouth?" Corporal Asquith said. "Whoever in the hell told you that had to be done?" the captain said.

DAY 2 OF *PLAN BUFFALO* RESUMED AT 0530 HOURS ON 24 MAY. The Wehrmacht hit the Allies hard on its west flank. German Mark VI Tiger tanks had no trouble with Allied infantry, and by the time Combat Command A Shermans reached the area, the Tigers had taken their leave. Fortunately, casualties to Allied infantry were light.

First Armored and 34th Infantry Divisions worked well together advancing toward Highway 7 at Velletri Gap, a vital push. Combat Command A then moved in further, using medium tanks to pound their way along. But by early afternoon resistance was heavy. The tough terrain between Highway 7 and the railroad lines made the going a slog. In the meantime, Combat Command B had better success than on day one. By nightfall it had captured 850 prisoners near the small hamlet of Cori.

Day 3 marked the successful completion of *Plan Buffalo*, as all northeastern objectives had been brought under Allied control. Cisterna was encircled. Aided by improving weather, air power was effective bombing and strafing enemy vehicles caught along roads northwest of Cori. The resulting carnage left hundreds of vehicles and as many as 15 Mark VI Tigers in heaps.

Meanwhile, word of success in *Operation Diadem* filtered in on 26 May. The battle for Monte Cassino Monastery, some sixty miles southeast of Anzio, had been a Herculean undertaking to clear the massive German defensive deep in Latin Valley. The abbey had been caught in the wrong place at the wrong time and had been overrun by the Germans in an effort to shore up their defense of Italy. From January to May the abbey had been subject to massive Allied bombings, but the rocked behemoth withstood all attempts to rid it of Nazis. *Operation Diadem* began in earnest 11 May using elements of U. S. 5th Army under Lt. General Mark Clark and the British 8th Army in a supporting role. This resulted in a massive Allied air attack on the German stronghold. With further help from Polish forces, the abbey was emptied of enemy but not before it the sacred place was left a pile of rubble. For Corporal Asquith, a practicing Catholic, this was a sad affair, discounting the Hitler apologist, Pope Pius XII who, like it or not, was still in charge of Asquith's religion. As the killing fields of this war tested Corporal Asquith's ideals of social justice and humanitarianism, his religious views also took a hit when considering the pope's questionable allegiance to the Nazis. It would shake the soldier's belief structure to its very foundations.

A 29 May the decision to move Combat Command A in position to attack along the Campoleone Railroad axis would prove pivotal. If this could be accomplished, it would allow for a breakout of truck-borne infantry. But a swarm of M-5 tanks got out ahead of infantry and in trouble with antitank fire. On the second day units attempted a breakthrough to the Alban Hills salient but that became as costly. Losses included 16 M-4 and seven M-5 tanks, and a half-dozen M-3 half-tracks. Troops dug in again.

Combat Command B had better luck by utilizing infantry, including 11th Armored Infantry Battalion. Campoleone Railroad had been taken and great strides in securing Highway 7 made. After four days of fighting the highway to Rome was within reach. Finally, on 2 June, Velletri Gap was in Allied hands. The *Hermann Goering Panzer Parachute Division* had given up and fled north, leading to the final phase of the campaign on 3-4 June.

––––––––––––––––

TRIUMPHANT WAS THE ENTRY TO THE *ETERNAL CITY*, 1ST Armored Division rolling in on 4 June 1944. Eleventh Armored Infantry would not be part of the roll-in, rather, it set up a command post near the suburb of Porta Maggiore located in southeast sector between Highways 6 and 7. On 5 and 6 June, Corporal Asquith and Whitey performed care and maintenance on their beat-up half-track. Loose equipment, most of it having been hastily piled into the bed, lay deep in dust. The half-track's left side, door to rear tail panel, was streaked with powder burns from a near-miss shelling on Albano Road; the right door forward to the front fender had suffered the effects of a mortar lob, dislodging the fender from the running board. The right front wheel and tire had been destroyed but quickly replaced; and the left front quarter panel had been repaired in haste using bailing wire. Heaps of 50 caliber machine gun casings littered the bed, marking a terrific assault directed at Wehrmacht infantry on Highway 7 near Velletri Gap.

Spud's wound was worse than originally thought, and he'd been shipped back to 95th Evacuation Hospital on Anzio beachhead; from there he would go out of theater. He had survived Anzio in body and—knowing the kid's spiritual resiliency—soul intact. Dago

Jimmy's body had been stacked like cordwood in a duce-and-a-half along with the other casualties and hauled off. Where he ended up was anybody's guess, although rumor had it a makeshift cemetery near Nettuno received many of the boys.

On 7 June orders came down for the Eleventh to spend a couple days of R and R in Porta Maggiore. Command issued GIs approximately $10 each in Italian Lira to spend the legal tender on goods and services in an effort to keep the area's economy afloat. Packets of prophylactics were included with the Lira. Corporal Asquith gave his rubbers to a member of his squad, never having an inclination he'd have a use for them. A few members of Dago Jimmy's squad mustered, eagerly conscripting Whitey to join them for one more extremely important mission: seek out, discover, and consume as many bottles of vino as they could in remembrance of the big lug from Missouri.

"C'mon, Quith," Whitey said. "You need to be with us to toast sarge."

"Sure, guys, sure." Asquith said. "Go on. I'll catch up with you in a minute or two."

But the corporal's plan was to remember the man with a long and quiet walk. With carbine slung over his shoulder, Asquith strolled alone towards the piazza in Porta Maggiore. A hot, noonday sun shone overhead. In spite of the heat, solders were required to remain in war dress, including steel pot helmets and loaded weapons. Hold-over snipers could not be ruled out. Most cafés in the piazza were full of GI's, their laps filled with Italian women, many bottles of vino weighing down the tables. Asquith did not feel like getting drunk. Instead, he simply relished the gaiety around him. In the distance he saw a yellow building. It stood out, not only for its unique European hue, but by the spray of bullet holes that danced across its façade. Standing in front of the building was a girl holding a bouquet of flowers. As the corporal approached, the girl stepped forward and presented him a nose gay of violets. Hypnotic were the girl's winsome eyes; her smile spoke volumes of appreciation. "Grazi," Asquith said. She was the most beautiful girl he had seen in his life.

PORTA MAGGIORE

7 June 1946

In July 1944 Corporal Asquith was promoted to the rank of sergeant. A new point system allowed many longer serving troops to be released from service. Whitey had gotten his orders to return stateside, and the two men promised to look each other up after the war. Before Whitey left, he and Asquith were able to toast Dago Jimmy with a glass of, appropriately enough, Dago Red. The bigger-than-life man had left an indelible mark on 11th Armored Infantry. Captain Shields had gotten wounded. He was hit by an artillery blast in the Po Valley offensive just days before the war ended but had survived with limbs and faculties intact. *Old Ironsides* had done its duty to the best of its ability—in spite of many fits and starts and wrong turns. The new leader, Major General Hobart R. Gay Jr.'s praiseworthy message to his men as they left the European theater was that of a job well done. He would lead the unit stateside.

Sergeant Darryl Asquith fingered the crimpled photograph of Caterina, its ink notation on back heavily smeared. Asquith had taken great care to guard it for the past two years. The photograph had accompanied him up the peninsula of Italy to the Arno River in August 1944; had been with him breaking through the Gothic Line in September 1944; ended the war with him in the Po Valley in May 1945; sailed with him on the ship back to New York in November 1945; rode with him on the troop train to Ft. Hood, Texas where 1st Armored Division deactivated in April 1946. The photograph took another train ride with him, this one to Albany, New York, then a bus ride to Schenectady where the returning soldier arrived

M. L. Graham

in May. Back home, he placed the photograph in the top drawer of his bedroom dresser, tucked in a memoir he had begun to write on the return-voyage to America.

The war veteran found his home much the way he'd left it. The modest house within St. Peter the Fisherman Parish had been purchased by his family shortly before his father's untimely death from tuberculosis in 1939. With war on the horizon, and needing to support herself and her son, Asquith's mother had gotten work at a naval tube assembly plant near Albany. Now at war's end, she had hired on with Smith Mercantile in Schenectady. Delighted in reuniting with her son, she had given him a warm welcome home party with uncles and aunts and scads of cousins—the majority of whom had missed the war due to advanced age or miscellaneous infirmary. The event proved an awkward moment for the retuned soldier. He, like many of his battle mates, was finding it difficult to acclimate after what they'd witnessed. Asquith's mother was deeply disappointed that her son did not want to go to Sunday Mass the weekend of his return home. The veteran soldier also refused to confess his sins to the parish priest.

Neither did he have any interest in the old '34 Ford jalopy. It was still parked in the garage with a dead battery. One day he folded open the car's bonnet and peered at the engine, but a flashback of his half-track caused him to quickly close the hood. He'd also run into Helen at the drugstore, she magnanimous on a farcical scale. She said there was so much to talk over since things had not worked out for her after the Dear John letter was posted, and oh how sorry she was about all the confusion "caused by this dastardly war." She invited Asquith to a dance and for some unknown reason he accepted. The dance was a total embarrassment as Helen openly sobbed on Asquith's shoulder the entire evening. She wanted to reconcile. But the veteran soldier had moved on; he'd not call on her again.

During the war Asquith had sent money home to help support his mother in her single-parent role. But with his mom working, she did not need the money and it had been placed in a savings account, the balance growing substantially. Asquith withdrew some funds, traveled to Albany where he booked passage on a Pan American Airways flight from New York City to Rome. From Rome he navigated his way to Porta Maggiore, arriving, ironically, two years to the day since he had met Caterina. The young man's excitement could hardly be contained. A day never passed in the final run of the war he didn't think of her. He wondered if she might not recognize him dressed in a civilian sports coat.

The only thing that had changed in the village was the condition of the buildings. Much progress had been made in repairing foundations and roof parapets. New stucco applications had received fresh coatings of paint. The piazza where his army pals had drunk to Dago Jimmy continued to bustle. The vegetable market and bakery were still there, and they were doing a brisk trade. He spotted Luigi Bianchi's winery. But Asquith did not tarry. His steps quickened; the palms of his hands were sweaty, and his heart raced the closer he got to Caterina's casa. There it was. The bullet holes had been patched and fresh yellow paint applied. He looked to the second-floor window. The red curtains where still in the kitchen window. Asquith entered the main door and took the steps, two at a time, to her flat. He knocked on the door, adjusting his sports coat and fluffing his shirt collar. There was no answer. Again, he knocked. He heard rustling and a baby cry. The door opened. Standing in the threshold was a woman— not Caterina—holding a baby appearing to be a year or so old. Asquith stole a quick peek into the flat. He could see a very large man seated at the table, his head bowed in stupor. A large wine jug rested on the table.

"Ah, Caterina?" Asquith asked.

"Che cosa, chi sei?" the woman asked. The baby began to scream.

"Caterina, Caterina?" again, Asquith queried.

"Caterina, non lo so…partire, partire," the woman said, gesturing Asquith to take his leave.

The man at the table snorted and looked over. "Cos'e questo, cos'e questo?"

"Sorry, so sorry," Asquith said, backing away.

Now what?

Darryl Asquith found himself in the piazza and at a loss for strategy. He then recalled the wine vendor, Luigi Bianchi, knew some English. He entered the shop. A young woman was stacking shelves with bottles of wine. "Posso aiutarla?" she asked.

"Si, ah, is Luigi here, oh how you say, posso parlare con, ah Luigi?" Asquith said.

"Si," the girl said and went to the back room.

Luigi appeared. He gave a curious look at the American, tipping his head, seemingly trying to place him. "Si?"

"Ah, Luigi, you speak Americano?"

"Si, a little, a little."

"Caterina, she lived at in the yellow casa, just down the street."

"Oh, you Americano here with the tanks," Luigi said.

"Yes, 1st Armored Division...June 1944."

"Si." The man looked towards his shoes.

"I want to find Caterina."

Luigi shook his head sympathetically. "Oh, Caterina, dear Caterina."

"She is not in the casa where she lived."

"Si, that is her sister and Caterina's little baby girl."

"What? Where's Caterina?"

Luigi's face was pinched with distress. "She, she died in child-birth over a year ago."

For Darryl Asquith the burden was great. He found it impossible to focus. Had she carried his child? The timing would be about right. What would he do now?

"Caterina is in the cemetery. Do you know where that is?" Luigi said.

"Yes, yes I know...I know where it is."

"Can I get you a glass of vino? I still appreciate what you and your boys did, rescuing my country from evil."

"Oh, no…no thank you…no wine." Asquith shook the man's hand. "Grazia, Luigi, grazia." As he walked away, he turned. "Did Caterina speak of the child's father?"

"Si," Luigi said, growing very ill at ease. "Well…Si. She said it was an Americano GI who lived in New York. And the man left to fight the Germans up north just after freeing our village of the Hun."

Asquith nodded and left the winery. He walked the short distance to the village cemetery where he found the bench he and Caterina had sat. He located the graves of her relatives then saw a new headstone. He walked closer. The marker read: *Caterina De Luca, born September 1, 1922, died March 12, 1945.* Asquith placed his head in his hands and wept. Then, as he slowly gained control of his emotions, he saw a nearby violet and stooped to pick it up. He smelled the flower and gently placed it on the grave. He then rose, stepped back and crossed himself.

Sunlight peeking through the trees cast afternoon shadows upon Asquith's face. A bird swooped down from a tree and lit upon Caterina's grave marker. It was the same variety of nutcracker he and Caterina had seen before. The black spotted bird let fly its pacey, rattling call. Asquith turned and quietly walked away. The bird flew off, climbing above the verdant canopy of trees winging in the direction of the Padiglione Woods.

M. L. Graham

SCHENECTADY, NEW YORK

7 June 1956

Que Sera Sera

In the afternoon shade of Sherwood Park's Cottonwood trees, Darryl Asquith sat with daughter Camille in his Ford convertible. The war veteran was smoking a cigarette while Camille licked an ice cream cone. Birds were flitting about. On the car's radio Doris Day was singing her just-released song, *Whatever Will Be Will Be*. One bird caught the man's eye. It was similar in appearance and call to the bird he had seen in Italy.

Camille looked her father's way and said, "Daddy, was mommy pretty?"

"Sweetie, your mommy was the prettiest woman in all of Italy."

"Will I be pretty as mommy?"

"You are already as pretty as mommy."

The girl smiled and licked her ice cream. The bird took flight, winged to the top of the tree line, dipped a little to the left, then continued on until it was out of sight.

THE PETER PAN

THE PETER PAN

1

Eldon Burdett took a moment's pause at the top step to study the blue Marlin hanging over the entranceway. It was natural to eye the trophy fish for it stood out, stranded—a fish out of water. Coated in a generation of urban dust, the fish had long lost its essence of prize, as had many who frequented the basement depths of the Peter Pan pool hall. In his early thirties, of slight build and a three-day beard, Eldon fashioned a twisted smile. "Hook, line, and sinker," he said to himself.

Down the man went entering the dimly lit chasm – a setting that might well have served as a cautionary theme in an evangelical sermon. Cirrostratus cigarette smoke hugged the ceiling of the place and schooners of beer, still fifty-five cents, fogged decision making – good judgment never having been keen with many of the place's clientele. A pair of ceiling fans struggled to move the stale air, yet a breeze of optimism welcomed those who hoped good gambling luck was just a cue stroke away. There, of course, was a bar. Marking much of its dark walnut surface was the inglorious marquee of permanent splash stains and an assortment of carved etchings—marrings whose authors lacked any historical significance. Bar stools were backless and wobbly and supported butts from mostly low socioeconomic walks – men who hoped their next billiard shot would be a winning one. Such optimistic air cohabitated with kitchen smells where a robust lunch crowd of blue-collar tradesmen wolfed down meat pies smothered in chili—and still under a buck.

A dozen regulation pool, snooker, and billiard tables were poised ready to test a player's hand-eye coordination. Players perfected cross-eyed banks to side pockets, paper-thin cut shots the length of the table, and proper management of shooting shape: the use of correct English that put deft spin on a cue ball, causing it to position for the next shot. It was an illusory term really – *correct English*. Nine-ball was the most popular game, and it was played by men whose thoughts of being snookered stirred primal fear. Here ringers pocketed wads of cash, and the not-so-lucky lost their paychecks.

A jukebox blasted out melancholic tunes that excused misspent youth and advanced a hope that losing was curable. Johnny Cash's ballad, *Sunday Morning Comin' Down,* was a hangover anthem. Little investment was put into anything new in the Peter Pan: table felt; repairs to leaky urinals; dynamic thinking. Any stable future beyond the next shot, like new tunes, made adapting difficult.

In the far corner, hidden by pony-walled partitions, was a collection of *Ballyhoo* pinball machines that paid off in cash for regular, trusted customers. The flashy contraptions featured gaudy lights and pinging flippers egged on by a harem of shapely female images displayed on a back-lit glass scoreboard. Wrist action keyed success to winning as players tested fundamental arithmetic with the dynamics of physics. Side-slapping skill might direct a chrome ball into the desired hole, hopefully without causing a tilt. A look at many of the players suggested life had already sent a fair share of tilts their way-- players seasoned in a mathematical gullibility that success had certainty. In the raw world of the Peter Pan the only certainty was the pinners were a profitable grubstake for the house, subscribing to the axiom, "There are many more losers in the world than winners."

Few women hung out in the Peter Pan, surely no storybook Wendys. Most of those who trickled in were damaged goods, and, like most of the men, their neglect was self-inflicted. Teased hair, overapplied eye makeup, and a come-on smile crafted many faces as carnivalesque. The strays lapped up alcohol. It flooded the voids of any dignity remaining—temporary fix as it was—and patterned an inability to trade up to those virtues capable in higher rungs of society. Some vagabonds looked to spin a quick trick—their version of nine-ball or *Ballyhoo*. For the men of the Peter Pan these women were affordable. The more beer the men drank, the better the women looked—the women counting on such transformation to get them through the night – with money in their handbag come morning.

For Eldon Burdett, observation had become a therapeutic tonic, easily transferable to his memoir writing. Frequently within earshot of the many tavern philosophers who offered unsolicited prose, he dismissed such unworkable wisdom; wisdom that became further attenuated the more schooners of beer the men drank. In spite of a capable intellect, Eldon struggled to stay upright in his unbalanced reality—it a product of his year in the Vietnam War. He was often found affixed to an end-of-the-bar stool nursing an 8-ounce bottle of Coca Cola. He didn't drink alcohol; he didn't shoot pool; he didn't smoke or do drugs. Such aberrant behavior in a place like this made him stand out. With the exception of Bev Lander, he had no close associations, keeping dialogue general and unexposed. He had gone to Vietnam in late 1967 just prior to the enemy's Tet offensive, an attack that drove the war bankrupt of its designed purpose. Sergeant Burdett met weekly with a Veterans Administration counselor, a man who frequently questioned why Eldon hung out at the Peter Pan. More than once, Eldon questioned himself about it. Moth-to-flame he figured.

In 1967 America was at a crossroads in the Southeast Asian conflict. As war death tallies intensified, blood splattering scenes oozed into American homes each night in living color as reported by the *CBS Evening News with Walter Cronkite*. Guilt began to haunt the collective consciousness, infiltrating college campuses and spilling over in protests on city streets. Yin/yang reared the head of a contrasting paradigm, as Americans watched protestors stick daisies in National Guardsmen's rifle barrels and burn draft cards. Still, more soldiers were shipped to the conflict. The collision of tensions juxtaposed the previous summer's euphemism—*The Summer of Love*. Now, even in the late summer of 1972, the war continued.

Pockets of the mentally ill hung out in the Peter Pan. Their numbers had grown exponentially with increases in population, a society at war and in economic downturn. Such conditions were exacerbated by an uptick in alcoholism and heroin addiction. Sammie Fremont was one. Tall and leggy and with a wobbly mental aptitude, Sammy had been unable to qualify for the military. Yet he stood soldierlike, ramrod in posture with a cigarette affixed to his lower lip as he chalked his pool cue, readying for his own kind of battle.

"So, Brooks, ya hear Wes wants back in?" Sammy said. Cigarette ash snow-flaked onto the floor, adding another layer of history to the place.

"Wunner if Walter will let him back in?" Brooks said.

Sammie's laughter detonated. It was an organic guffaw, the conspicuous kind that caused heads to turn. Eldon mused it *nuthouse laughter*—laughter which incited more laughter, and the kind he heard on occasion in post-discharge confinement at the mental wing of West Port Veterans Medical Center.

"Boy, he was sure pissed when he left," Sammie said as he approached the end of the pool table. His pool stick slammed the cue ball into the rack, scattering object balls helter-skelter. None found their way to a pocket. "Damn, you're go, Brooks."

Brooks stepped forward and bent himself at the waste, surveying the shot's angle. He straightened up, cued the stick and took a stroke, pocketing the yellow one ball in the corner. It made a soft plop in the leather pouch and caused Sammie's laughter to reengage. Brooks chalked his cue and moved around the table. His movement was more a float, reminiscent of Jackie Gleason's in *The Hustler*. He lined up the blue two and shot, sending it to an awaiting side pocket. The run was on.

Sammie retreated to a stool beside Eldon and mashed his nearly spent Marlborough into an ashtray. Compulsiveness caused the man to work the flip-top box for another cigarette. A flash from his chrome lighter made flame and flirted with the tip of the cigarette producing a plume of smoke. The process climaxed with a metallic snap as the lighter was closed and pushed back into his Levi pocket.

"Better rest awhile, Sammie," Eldon said, "looks like a run."

"Hell, I busted 'em balls up too good." Sammie punctuated his statement with another volley of laughter. "Leave me some, Brooks." Brooks winked, chalked his cue and continued.

Eldon knew Wes equaled trouble. A rawboned ex-convict with battle-hardened face, Wes Richardson had a malicious sense of self-importance; and when coupled with a dangerous knack for recidivism, the man was capable of headline news. It had only been three months since his release from the *point-of-the-mountain*—the felons nickname for the state's penitentiary. Two weeks before, Wes, drunk, had been 86'd by Walter, the Peter Pan's bartender. Wes left in a huff, and many thought he'd return with a pistol—a pistol the man had no right to possess. Fortunately, as rumor had it, Wes suffered a full-of-drunk mental lapse for retaliation and landed around the block at the Havana Club.

Brooks finally missed a shot and gave Sammie a turn. Eldon made small talk with the bartender. A longtime employee of the Peter Pan, Walter had been bartender for 25 years through two changes in management, the cop's raid in late 1965 and small fire, accidentally set by a night cook who left a lit cigarette on an oiled chopping block. It rolled into the stove's grease tray, smoldered, and filled the place with smoke but caused no structural damage.

"Bev was in for lunch earlier," Walter said.

Eldon shook his head. "I know, damn, I forgot we were meeting for lunch."

"She seemed fine. Had a ham and Swiss on rye then went back to work."

"I'll see her later. She's used to me standing her up...has a lot of tolerance."

The late afternoon lull offered a sense of calm. Pool halls and taverns were like that. The lunch rush had cleared, and the after-work crowd would soon arrive. Walter's natural barkeep reflex was to towel the counter and recheck his stock.

"So, Sammie says Wes…

"No, he ain't," Walter shot, interrupting Eldon. "I said the bastard was finished here and I don't mean maybe."

Eldon sat back, cocked his leg, and took a sip of Coke. Walter tossed the bar rag down, set his hands akimbo, defiant, and rocked his upper torso back and forth.

"You know his damned probation officer came in here. Now I ain't a stoolie—never have been even if it flies in the face of a job like this. I ain't lookin' for no trouble, but there comes a time when a man's gotta take a stand. When that shithead pushed Cripple Louie over…goddamned crutches on the floor and his metal leg braces making a hell of a thrash…and Wes then staggers around Lou heading to the toilet and laughing. What kind of a man pushes Cripple Louie down? I told Mrs. Butler, I don't know if you knew this, but she just took over running the place. Anyway, I told her it just ain't right…that shit ass picking on Cripple Louie. I'm not sure she saw it my way. You know, she…"

"Hey!" Sammie exclaimed. "Did you see that shot, Walter?" The blue four ball had ricocheted into the yellow stripped nine, sending the money ball into the side pocket. Laughter punctuated Sammie's outburst.

Walter's focus never left Eldon. "I tell you, if he walks in here, I'm calling the law…and that's that." Walter walked away in a huff, stopped at the TV located center bar above the mirror and flicked the dial. He took a step back, focusing on the screen. "And if all that ain't bad enough, good god, now look who's on the television." The four-o clock news featured clips from a press conference on embattled Vice President Spiro Agnew. "And that son-of-a-bitch is as guilty as sin if you ask me. I call him *Spiral Ga-new*." Walter hooted, liking his joke. "What the hell Tricky Dick ever saw in that shyster, I'll never know."

Eldon smiled, shifted on his stool and turned his focus to the TV as he nursed his cola. Walter toweled the bar while Brooks racked a new game. Sammie fired up another Marlborough. At a quarter of five the Peter Pan's door swung wide, allowing in a slant of sunlight and push of August heat. The jangling of metal banging against metal was heard, followed by the trundle of shuffling feet. The feet moved in rhythmic lurches, interspersed with a jarring clank: *bang, slide...bang, slide...bang, slide...*

Cripple Louie was a man whose very name was caught awkwardly between an era of cruel bluntness and political correctness. Louie was a fixture in the Peter Pan. His club foot, twisted elbow, and elongated face cast the unfortunate fellow in ghoulish persona. When on the move, his slip-sliding movement might have snagged him a part in a Lon Cheney movie, leaving his drawn, macabre face completely out of any screen shot potential. For Eldon, poor Louie begged the question, could there possibly be an all-loving God? Louie slid past Brooks and Sammie and stowed his bariatric crutches in a lean-to against the backstop of the bar. He positioned his good elbow on the countertop and hoisted himself onto a stool two down from Eldon. Walter moved back down the bar and assumed his akimbo posture.

"Hey, Walter," Louie stuttered. "Ah have a Fresca?"

"Hot one for you out there, Lou?" Walter said as he reached into the undercounter cooler. He pulled out the soda, removed its cap with a church key tethered to the bar by a cord, grabbed a tumbler, iced it, and slid the order in front of Louie.

"Kind of sultry today, Walter," Louie said, looking up at the TV. "What's on?"

President Richard Nixon's mug, highlighted by its signature five-o'clock shadow and *not guilty* sneer, filled the screen. At a press conference, he was heaping praise upon his Vice President—a man who had been accused of accepting kickbacks from contractors during his tenure as Maryland governor.

Walter had no reply, instead shook his head, took up his rumpled towel and moved further down the bar.

"You believe the president, Louie?" Eldon asked. The two laughed and nursed their drinks.

It was a little past five when Bev Lander appeared. A woman in her mid-forties wearing a light summer red dress, she carried a large handbag and wore a sun hat, an expansive hat set at a stylish tilt that proffered feminine chic. A discerning eye might have noted the questionable affordability of her second-hand dress but few such eyes in the Peter Pan qualified. Her easy attractiveness threw her in contrast with the limited run of other female clientele who happened upon the place. She walked to Eldon's end of the bar, pushed her purse onto the counter and slithered atop the stool between Eldon and Louie. In an awkward, sideways gesture, Louie's attempt to make more room for her resulted in Bev's, "That's okay, Louie, I have room."

"Hi, Bev," Eldon said, a distance to his voice. "How'd the day go?"

"Oh, same old cheeps and creeps. You know with the liquor store next to the emporium, they get all boozed up and come in and want to talk."

Removal of the hat revealed light brown hair, clean and shiny. Her lips had a natural upward-bend, and a light shade of lipstick politely complemented her clear complexion.

"Sorry about lunch," Eldon said, his Coke bottle now empty. "I got carried away."

"Well, you're a writer…isn't that something a writer does…gets carried away?" It was not a harsh statement.

The aura of Bev's smile transformed her face into the special woman Eldon feared he was growing too serious over.

"You want something?"

"No. Just ice water. God it's muggy today." She looked up the bar. "Walter, can I just get a glass of ice water please?"

Walter arrived with a tumbler. "Tough day?" he asked.

"God, don't get me started."

The sharp crack of billiard balls reminded all this was still a pool hall. Walter smiled and walked off. Eldon and Bev made small talk. It was not long before the door began letting in a flood of the after-work crowd on this thirsty Thursday. In no time a quarter found the jukebox slot, queuing up a set of three songs. Music strong-armed the evening TV news, silencing even the President of the United States. Louie drank his Fresca. Sammie lined up the angle of his next shot. Walter was down the bar, content to be busy. A snappy uptick in business had replaced the late afternoon doldrums, as cash register dinging promised to make Mrs. Butler a happy entrepreneur.

2

Eldon sat with Bev on the number 10 Popular Grove bus as it pushed along in rush hour traffic. They had been lucky to find a seat. Commuters were packed like cord wood, the press of standing ridership holding onto poles and ceiling straps and swaying to and fro. Subtle pongs of mostly spent perfume and deodorant hovered in the air of the westbound bus—six PM scents much faded from eastbound's morning bus. Eldon contemplated the close-quartered journey. Jostling crowds made him nervous, a byproduct of his Vietnam War experience. Most people wore vapid, commuter faces. Eldon assumed many of the passengers had completed wearisome, daylong tasks and now an evening of vague domestic routines awaited: feeding the kids, walking the dog, sit-coms, and more political corruption reports on Spiro Agnew. It was a life Eldon had no interest in, even if conventional employment might allow him to live more comfortably. He'd proven one could get by cheaply with discipline and forfeiture of domestic entrapments like home mortgages, car loans, and conventional family ways. The monthly stipend he received from the VA helped.

At each stop a blast of hot desert air infiltrated the bus, adding an invisible dose of smelly urban particulates to the mix. With resumption of travel, a sneeze from the bus's hydraulics seemed to strain the air-conditioning system, making the sultry worse. Eldon redirected his attention to Bev's dress and exposed knee. Her dress was bright and sunny, and he disregarded a couple frays in its hemline. He had been with her the day she bought it at the second-hand emporium—before she had gone to work there. It was a happy dress and affordable. Bev had been blue that day and needed something to cheer her up. She'd just learned her son had been arrested again on drug charges in Denver, and it might mean prison time.

Bev spoke of her son with a soft nostalgia. Once, confiding in Eldon of the child's history, she confessed marijuana had led to cocaine use, then to pregnancy. It plunged her naturally into a funk, bottom of the heap with no means up and out. In a period of fleeting permanence, the father was a question mark and, as her belly expanded, it soon passed the point of no return for a dangerous, back-alley abortion. Plus, she could not bring herself to sharp instruments violating her body. She'd carry the child to term then give it up. But when holding the infant on her belly after birth, this was no longer an option. She had no inherent child-raising skills— her mother long, long gone and other next of kin in different time zones. She did the best she could with what little she had. Struggle was a fixture in her life until the boy came of age. Now he was on his own but still the struggle persisted.

Eldon held Bev's hand, still feeling a little self-conscious with the gesture. He lacked the gentle disposition for touching, perhaps owing to a masculine, blue-collar upbringing. The army reinforced the trait, hardening him into a tactical team member with strong mission achievement objectives. Prior to his time in the military Eldon had a go at love. It had a common storyline: the girl waiting back home. But it didn't work out. After his discharge and meeting Bev, he made strides to get over it. From their first meeting Eldon found Bev easy with conversation, and they soon discovered each had similar socioeconomic history. Convenient sexuality blossomed – blossoming, perhaps, as much as two people with heavy mind baggage might achieve – at least for now.

Discovering that his love making could obey the pleasures of a slow, tactile journey, Eldon learned Bev appreciated such rare patience. For her it might have been a relief from the usual slam and bam process most men in her life were known by. Bev's love making appetite was whetted with a nip of bourbon. Eldon could not always afford a pint for Bev but was rarely without a trio of mini-bottles, the kind now in his pocket. His free hand pressed at them, an edgy habit. The stockpile made dull jingles—sounds mostly absorbed by the hubbub of the commute. His eyes moved easily up from Bev's knee to her breasts and eventually her face; her reaction was a silent smile telegraphing a healthy inclination for the evening ahead. Two drinks normally set her at ease from a hard day of retailing and being on her feet; a third was reserved for after—the cleanup; dressing; the quiescent recapitulation. Bev would then leave for her nearby apartment and Eldon would go back to the typewriter and reengage his dragons as per the VA counselor's recommendations.

Eldon lived on Pierpont Avenue in a pre-World War Two bungalow duplex. It was a two-front-door affair, each side sharing a common bearing wall, a wall that leaked cooking smells, impatient shouts and spin-off arguments. Thank god there were no wailing kids. Eldon had no TV and only a portable radio on the kitchen table usually tuned to an easy-listening station with its less jarring genre of contemporary music. Eldon banged out his stories on a Royal manual, forsaking the newfangled IBM Selectric, the self-correcting gizmos whirring in newspaper rooms and upscale offices. A great-aunt had left him a few shackles and with the 60% Veteran's Administration disability, he was by no means a wealthy man but got by. Yet, even with those tolerable domestics, it could not prevent incursive dreams from messing up his wellbeing.

Earlier in the Peter Pan with the TV spilling rhetoric on Spiro Agnew, Eldon was reminded of the political deceit Vietnam had become. He never talked of his participation in the misbegotten debacle at the pool hall. But keeping it inside made him vulnerable when asleep. Reoccurring dreams often spring-loaded ambushes, and they snapped at him like the sharp bamboo points of hidden booby-traps. Hours of VA counseling had determined the best way to deal with his dragons was to bring them into the open. Eldon liked Doug, his counselor. The man's offerings were sincere in spite of the fact he was heavily steeped in the didactics of textbooks and light on raw war experiences—he had escaped Vietnam as a member of the Army National Guard. Still, he couldn't fault the man for being a reservist and missing the war. Fate's sketches are oftentimes an oblique artform.

After Eldon and Bev had intimacy and the front door closed behind her, Eldon took up his position before the typewriter at the kitchen table. He rolled in a fresh sheet of paper and resumed his two-fingered tapping:

My military hitch took place during an era of bipolarity in America's geopolitics: the war in Southeast Asia roiled during the Summer of Love in 1967. Such dichotomy flowed in our collective veins then. There had been a woman in my life, a first true love whom I had known since high school. Our relationship developed with the speed and fury of the Johnny Cash and June Carter song, "Jackson," often played on the Peter Pan's jukebox. Despite such pepper-spout heat, our time together was extinguished as many relationships are with a breakdown of spirit and trust. But dreams of Carmen hung on with frustrating regularity until she was replaced by another woman. Her name was Annie.

Annie was the affectionate name for the M-14 rifle I met at basic training in Ft. Ord, California. As Drill Sergeant Sandoval extolled: "A gun is what you've got between your legs. The M-14 is a weapon. Treat Annie with care and kindness, and she'll look after you." My platoon courted Annie for three weeks of double-time marches to and from the gunnery range. Such marches trudged along in choking dust, dampened only by the platoon's collective outpour of phlegmy spit and profuse sweating.

We became intimate with Annie, undressing her every evening. We learned Annie's inner workings, her spring loadings, her trigger housing; we palpated her ammo clip orifice and polished her barrel; we stroked her fine, smooth wooden butt. And as with any woman, she had to be treated with kid gloves least she break your heart (cause a miss-fire that might get you sent stateside in a rubber body bag).

Annie's mood was influenced by the number of elevation clicks and side windage adjustments controlled by two knobs near the barrel's aft gun-site. These controls were the main factors in achieving a tight shot group on a paper target. After a round was fired, we adjusted elevation and windage controls for accuracy and fired once more. We learned to gently squeeze off a round, for a jerk of the trigger would undoubtedly make you miss your mark. And if you missed your mark, the enemy was apt not to miss his.

Ft. Ord's firing range had its own peculiar dichotomy: surrounding the perimeter were thickets of ice plant succulents with luscious green leaves and fragrant, yellow flowers. It offered resolute beauty amongst the flashes and explosions, impact concussions, and overpowering stench from the weapon's gun grease. War facilitates such contradiction.

We spent an intense three weeks courting Annie. One in our ranks, Gardner, slept with her. But most of us developed a wary intimacy, for we meant to break off the relationship when we returned to "the world" – unless orders for Vietnam arrived. I had never been into guns—I was not a hunter or carnival midway participant. I was simply trying to muddle my way through the army experience. When I ended up qualifying as a "sharpshooter," below "expert" yet higher than basic "marksman," I was extremely proud of myself. Yet I knew that when I put Annie back in the Ft. Ord armory for the last time, I had no plans of ever pulling her trigger again. But with the "best laid plans of mice and men," orders arrived for Vietnam. After a brief furlough I mustered through California's Travis Air Force Base, next stop South Vietnam.

Merciless heat, soaking, sticky, all-encompassing, pushed aboard the World Airways 727 jet the moment its aft stairway door opened. Then came the stench of musty decay.

"Welcome to South Vietnam, ladies and gentlemen," a voice announced via the airliner's intercom. "As you file off the plane, collect your duffle bag then muster on the tarmac single file with orders in hand. You will then be assigned the appropriate bus that will take you to reception halls. From there you'll all be processed to your final destinations. Look alive, people, look alive. You are in an active combat zone."

Sweat puddled under the felt-wool cunt cap atop my head making for an efficient dam. We newly arriving grunts had been briefed on Southeast Asian climes at Travis and were happy to have been allowed Class B light kaki shirts and matching trousers for transit. But the wool garrison cap had once again brought to the fore a well-known military axiom: "There is a right way; there is a wrong way; and there is a military way of doing things." The story is as old as General Pershing's horse battalions. And further complicating the Army's formula was the continued blitzkrieg of newspaper reports on growing tensions the unpopular war was having upon American society. Even before I had put a boot on the ground, I longed for a return to "the world."

By the time I arrived in Tan Son Nhut I had achieved E-5 Sergeant and made a tolerable transition to Delta company. It was not long before I learned the yearlong commitment would bring a number of replacement leaders, 2nd lieutenants mostly. But one in particular stood out. His name was Lt. Grant Williams; and what a shining young man he was...

Eldon ceased typing when the radio station transitioned to headline news. He rose, switched the dial off and headed to bed. As he climbed in, Bev's lingering scent was still detected.

Next morning, coffee cup in easy reach, Eldon again loaded the Royal with paper:

...Lt. Williams was a quick learner, a must if any reasonable life expectance was to be had. I got to know him better over rum and Coke at the rear—if "the rear" was possible after the Tet Offensive in January 1968. I learned he hailed from Washington state, the Palouse, high on the eastern plains. He had no misconceptions of war and confided in me he had learned the first week in country to toss out much of the bullshit taught at Officer's Candidate School (OCS) and the skimpy pre-deployment muster. I told him that we could all benefit from that way of thinking and, perhaps, live a little longer.

War had been a family affair. His dad had been with General Mark Clark in the early Rommel chase across North Africa in 1943. This pursuit resulted in failure for Rommel's men when he was recalled for counseling to Adolf Hitler's Wolf's Schanze retreat deep in the Gierloz Forest of Poland in the middle of the African campaign. There, the talented general would learn of his role in a new theater of war called Normandy. It would not end well for this man called the Desert Fox. Like the Fuhrer, suicide would be fate's common denominator.

"This war ain't like that one," I said with a stiff chuckle. "My father had been with the 1st Armored Division in the African theater as well...there and at Anzio's baptism by fire. My dad was really against my joining the army...said the Vietnam War made no sense to him."

"I know, Sarge, I know, it's my past...family stuff," he said. "My dad did not want me involved in this war either...said the same thing."

I laughed. "Well lieutenant, and here we are...maybe the old guard was on to something."

M. L. Graham

Lt. Williams then revealed the powers that be were aiming to up our platoon's scheduling in sending us out into the jungle for the same-old, same-old: get Charlie.

"We need to be less efficient on the next mission," I suggested. "The brass keeps notes and will be quick to redeploy if a squad becomes too efficient. Hell, I'd learned that way of thinking in basic training at Ft. Ord."

"Orders are orders."

"Suppose so," I said, taking a pull of strong dark rum and thinking about Ft. Ord.

Yes, Ft. Ord. It was there in week two of boot camp I got a real lesson in life when a small package arrived from Carmen. Enclosed was the engagement ring I had given her the night before I shipped out. Accompanying the box was a dear John letter. This second week of basic had been impactful—a turning point. For it was also that week I learned the riddle of the quick and the dead. The "quick and the dead"—it had a brassy ring, didn't it?

Lieutenant Williams had arrived in country mid-December 1967, six weeks before Tet twisted the war into an ugly tangle of deceit. Homogenized by OCS's "Shake and Bake" ineffectual World War II strategies, the good lieutenant's dismal naivety was further heightened by his teasing resemblance of the kid on the label of a Sunny Jim jelly jar. On the day he arrived in the Nam, such a fresh, farmhand look could not disguise the man's body odor. This nervousness was a common smell for new replacements—a combination of the insufferable wet heat of Southeast Asia and the raw stench of stress that rose from a first combat assignment. Opening firefights usually dissolved the smell, melding it with the rest of the platoon's musty aroma. Depending on the officer's chops under fire, acceptance to the platoon was established then, thus securing future missions.

It was a rough task for shavetails. Second Lieutenants were expected to be instant leaders of men inside a war theater that was quickly racing through its inventories of body bags. These rubbery sacks were a chopper's delivery behind the firefight, normally packed conveniently with arriving litters—litters reserved for those lucky to still have a pulse.

Lt. Williams inherited a company of soiled and disgruntled grunts who had forgotten quick as they could their last leader—the normal turn of the screw. It was pretty much assured the first patrol would run the new leader to ground, especially with all the splattering's of bullshit taught at OCS. Laughable really. Officers and non-commissioned personnel (Non-Coms) were all affected by the ill-preparedness bullshit, standard GI issue for newcomers to the Nam. To see the look on new faces when they saw their first splashes of blood red on green undergrowth; the body gore; the smell of gun grease and rubbing alcohol. There was no letting go of the scene, quickly retrievable in sleep when the patrol was thought to be ended.

"This fucking piece of shit's jammed again," was common jungle call in those early days, expletives shouted above the metallic clunking of a malfunctioning M-16 rifle. "The Pentagon's highly aware of the jamming problem men, and they're working on a fix," came the bullshit from Military Assistance Command (MAC). In the meantime, for many, trust was placed in God. It was hoped He might be near, perhaps in a low crouch beside true believers.

Late 1967 and early 1968 became a revolving door of replacements in the struggle to become short: going under six weeks left in country. Many new arrivals notched remaining day counts on the calendar, but the habit fell away. Some resumed counting when they hit 60 days to go, but that could be dangerous. Renshaw got his with 59 left. Shit! The bastards!

I remember it was a Monday ("Monday, Monday...can't trust that day..." as the song went). We were on search and destroy and pinned down in mud and rice paddies with radio-man Jensen barking for air cover. That's when PFC Phillips got it. He was hit in the upper arm, the slug ricocheting off his humerus and somehow ending up lodged near the elbow. He was a high-strung Nebraska boy, the most dangerous personality to have in the Nam—always blathering. He got on most of our nerves…had the archetypical long-haired girl awaiting him back home, but the more seasoned of us knew that was horseshit. She was fucking 4F Jimmy now; it had happened to a lot of us and, in fact, now in the mud and humidity, that kind of thinking offered up an arrogance that pissed us off to the point it made us survivors. This is what the war did to you.

Absorbed in the whoomph, whoomph of an "Air Mobile" Huey evac helicopter setting down a grenade's toss from where radioman Jensen shadowed Lt. Williams, I helped the corpsman with Phillips. We got him aboard the chopper then ran in a stoop back to our starting place. The bird lifted away. Lucky kid. He was only winged. But he'd be going home now and would learn the news about Jimmy and his gal. The double whammy of the Nam.

Lt. Williams proved a quick learner. Later in the week we were pinned down by what we thought was a squad of Vietcong. Unmuzzled now, my squad had matriculated to monsters—monsters fighting monsters—retreat never ran in our head. It was too late now. We could never return to "the world" with the same naiveite we'd left home with.

Lt. Williams motioned his paw for us to stay down. Wapp, Wapp! It appeared to be a sniper. We pressed belly-hard in the mud. Then, Christ, for some inexplicable reason my mind flashed on the last time I saw Carmen when leaving for boot camp. Her eyes were full of despair, tears streaming down her cheeks. I tasted her tears in our last embrace…so bitter. Now a song, a damned song, "Nights in White Satin" by the

Moody Blues flew into my head. How could this be happen-ing in the middle of a goddamned firefight in the sticky rice paddies of Southeast Asia? My life flashed before my eyes. I felt my extremities and as far as I could tell I'd not been hit. But this song?

Finally, thank God, a rattle of 50-caliber from PFC Bor-delon's weapon sent a fusillade of destruction towards the distant underbrush, tearing to shreds the reverie of Carmen and stopping the music. Fire was directed to where it was thought a sniper was positioned. The shots kicked up water and mud and ripped apart branches of mangrove. When the firing ended there was no movement...nothing. Maybe he was dead; maybe he'd gotten away; maybe he was never there...

Eldon pulled the 20-pound bond paper from the typewriter and set it in a tray with the others. He stuck a fresh sheet on the carriage, rotated the roller, positioned his index fingers, and began to once more pound the keys. It was therapy, he reminded himself.

3

With impish grin and eyes in search of one's vulnerability, Wes Richardson's face might have been found on post office bulletin boards. Yet, obsessed with such facial curiosity, one could have missed the virtuosity of how the man's index finger met his thumb when cueing a pool stick. It was a finesse that, when allied with an uncanny eye for angles, resulted in the efficient separation of a pool shark from his money. Encounters with Wes assured a losing proposition. But this was the Peter Pan, and a sizable flock of pigeons queued at the chance to take him on. In more innocent walks it might have been natural for men to simply flee, cut their losses, what with his ruffian reputation and deft skill at nine-ball. But true gambling men failed to understand the velocity of money. It was more than a fundamental Economics 101 concept. A draft beer promised better return on investment, although it too had its short-run downside with a frequent need to piss. No, men of the Peter Pan were easily snookered when trying to make a fast buck. Such gullibility stuck to them like city street smells.

Wes had been let back into the Peter Pan by a raw talent for manipulating the powers that be: Mrs. Butler. The new owner of the establishment was a seasoned manipulator in her own right, so this relationship promised future fireworks. Walter had informed Eldon Mrs. Butler's divorce had given her the ramshackle pool hall, and she was determined to make a go of it. Eldon estimated the woman to be in her fifties, so association with a thirty-five-year-old rough and ready ex-convict promised to be a page-turner story. Mrs. Butler sat at a table going over paperwork with her two daughters. Barely old enough to be in the place, they were pretty-faced cherubs dressed in scant clothing that revealed rippling cleavage—cleavage which had gained Wes's attention as he leaned on his pool cue awaiting his turn at the table.

Walter was not impressed with the developing scene. "Chrissakes she's drinkin' wine coolers and smoking 'em damned Old Golds," he whispered to Eldon, who, as usual, was holding up the far end of the bar. "Who the hell still smokes Old Golds?" Walter toweled the bar and shook his head, keeping a cautious eye on his new boss.

Eldon had known such men. Wes's suggestive glances directed at the young prey was most likely a tight-roping plot for two-timing his new girlfriend for her daughters. He'd be very capable of something like that. Eldon figured Wes had visions of a frolic twosome with the fleshy girls—hell, maybe an adventurous threesome should Mrs. Butler's wellsprings dictate and join them should a stout bed capable of such an event be found.

Saturday had filled the Peter Pan with pool hustler wannabes: a weekend species of truck drivers and office clerks and medical students from the university—all very low on the serious punter's food chain. Bev was working some overtime at the emporium, so the daughters were the only women beside Mrs. Butler in the place; and their mother seemed to be enjoying the attention her girls were getting from this bastion of horny males. In between deep drags on her cigarette, Mrs. Butler reviewed stacks of invoices with them while one of the girls fingered a battery-operated calculator.

A fair-sized audience had gathered in the spectator area, a cordoned-off box seating section near the pool table where Wes held court. A stack of dollar bills rested at one end of the table. Wes's turn came to shoot. A puffing cigarette in his mouth sent smoke across his face but appeared to have no ill effect on his vision. As he had called it, the green six ball was sent ricocheting into the yellow-striped nine, and the nifty combination shot caused the money ball to plop into a side pocket. Spectators clapped. Wes reached for the stack of bills, careful to show off a black widow spider tattooed on his forearm.

"Thank you kindly," he said, stuffing the bills into his shirt pocket. Next."

Cripple Louie had just jangled through the door, spotted Wes, then redirected his wobbly skid towards the far end of the bar where Eldon sat. Walter had Louie's Fresca before Louie could stabilize himself on a stool. The usual pleasantries between barkeep and patron were exchanged. A momentary lull in the big game-of-the-day was declared when Wes announced for all to hear he needed to take a leak. He set off for the latrine, a route that would make a close encounter with Louie and Eldon.

"So, how's life treating you, little man?" Wes said. Louie ignored the salutation. Wes then eyed Eldon. "See you're on your perch, chirp. Chirp, chirp." Wes's chuckle was the vituperative kind. "How's the cola? You need a little rum in that, pal. Might make a man out of ya...make you a little more talkative too."

Wes moved on, using both hands to push against the men's room door, disappearing behind it. Louie stared down his bottle of Fresca. Walter watched Eldon, Eldon's lips on the verge of a smirk. A few moments passed. Eldon's facial reflection of himself in the bar mirror transitioned to a more congenial, albeit schizophrenic smile. He took a sip of Coke. This was not the time nor place to deal with such a goon.

For Eldon, a reckoning was at hand. Conflict had been building in his head for some time even as the VA counselor seemed to have been pleased with his patient's progress along the narrow path to recovery. Doug suggested the purpose of the writing was to give Eldon the opportunity to first confront, then break the back of his depression. It was a peculiar juxtaposition, Eldon thought—using something depressing to quell depression. Strange were the workings in the field of psychology. Others under the counselor's care had made their way out of the tunnel with artwork—the old adage beauty being in the eyes of the beholder. Yet, Doug was a realistic counselor and had confessed to Eldon he may never get completely over it—slay the dragon as he often said. At least Eldon's financial lifestyle was stable, unlike many others under the counselor's care. As long as he did not resume drinking or get back into drugs, he'd have a chance at wellness.

Doug also encouraged Eldon's love interest. Bev had been discussed on a number of occasions even though such two-party dynamics might add a complicated mix to the healing process. Still, Bev was a ray of sunshine in Eldon's life, he could not deny. She offered the injured soldier felicity, understanding, caring, and kindness. But depression's strong undertows complicated love's interpersonal mechanics. It had always been easier to return to a quiet, self-centered observance in the Peter Pan. It had become his church, peculiar as it was with its uncertain deity and twisted form of worship. Vietnam had taught him that the conventual Judeo-Christian world was an illusion. The God that might have lurked in mangroves and rice paddies had simply been a no-show on too many occasions. Many of Eldon's wounded mates had cried out for Him in their time of need. For the curmudgeon sergeant, God had been shipped out of the Nam in His own body bag?

Wes emerged from the latrine purposefully zipping up his trousers so the trio of females could witness the act. The rogue's gallery seemed happy the king of nine-ball had returned to his court. The next combatant stepped forward, laid down some bills, and chalked his cue tip. It was nearing noon now and palpable was the Peter Pan's entrepreneurial buzz. The kitchen was hopping, sending out trays of meat pies and hot pastramis. All the pool tables were in use as were most of the pinball machines. Such commerce promised a good gate for Mrs. Butler. When a quarter deposited in the jukebox produced the Association's *Along Comes Mary*, Eldon abandoned his stool. No woman with such a Biblical name hand any business being in this testosteronal gathering of men in rut. He headed for the exit with no interest in making eye contact with Wes.

Back home now and at the typewriter, Eldon pushed a few keystrokes but could not get back in the flow. He needed a walk. But what he really wanted was Bev, and that was not going to happen until Monday evening.

4

Bev's tongue tasted of bourbon, her breath sweet with hints of liquor from the just-finished mini-bottle. From the other room, barely audible, the radio aired *Loneliness Remembers What Happiness Forgets,* a song by Liz Damon and the Orient Express. Its soft harmonization added a coziness to the dim lighting of Eldon's tidy bedroom. Bev was especially passionate this night. She had told Eldon on the bus ride the girl who took up with her son in Denver had been communicating with Bev and was remaining by his side. She had sought state medical help, and the agency agreed to take on his case. The relationship between mother and son that had crashed a year before might now be reconstructed. Bev wanted that more than anything.

The two had fallen into their likeable positions, she atop him, her flowing hair dancing across his chest. Droplets of tears moistened her eyes, something Eldon had come to expect with Bev's love making—unknown if such emotions were raw excitement or wistful foreboding. Bev had had many lovers in her life, Eldon only one. But past relations were never discussed. In this stage of their lives the couple had become weary of demanding allegiance. Now, a couple times a week, the two immersed themselves in an acceptable sexual arrangement – one without remorse.

Finished now, the two administered a deep embrace before Bev fell away to the side of the bed. Both lie on their backs in silence, looking at the ceiling and catching their breath.

"I didn't mean to go on so much about Erik earlier but do need to ground," Bev said.

"No, no, Bev," Eldon said, turning to his side and kissing her forehead. "It is good news…the best. I told you if you need to go to Denver to be with him, I can help with the expense."

"I don't need to do that. He needs to deal with his own life now… and the girl with him seems helpful when I talked with her. "

"Well, the offer can still be managed so keep that in mind."

"You've mentioned this Doug from the veterans place has been helpful for you. I'm hoping counseling will help Erik. And you keep so busy with your writing." Bev pinched Eldon's side. "But you spend too dammed much time in that run-down pool hall, I know that."

"It's a good place for me to think. There are some good people there."

"I know. I love little Louie. He's such a nice man caught up in a cruel world. I hate it how some people tease him."

"Yeah, I really like Walter too. He's seen a lot in his life…knows how to judge people. Not sure if you know that Wes guy Walter 86'd."

"Oh, that bastard? I've seen him a time or two. He always undresses me with his eyes as I walk past him. What a creep."

"Well, he got let back in."

"How'd that happen?"

"The new owner. Walter told me about it Saturday.

"I didn't know there was a new owner."

"Yeah. It's a woman."

Bev laughed. "Good, a woman. Maybe she can clean the place up. But letting that guy back in, well, she's off to a bad start."

Eldon sighed. "What somebody needs to do is clean the guy's clock."

Bev swung her legs across the side of the bed and sat up. "I need to clean up." She rose and moved to the bathroom to take a shower. Eldon got up, dressed and went to the parlor where he took a seat on the davenport. Bev soon emerged, still naked, standing before Eldon and toweling her hair. Her well-proportioned body was lean and sensual, and her gentle smile had the capacity to dissolve Eldon's deep anxieties—if he'd let it.

After Bev dressed, she found the remaining mini-bottle. She twisted the cap and had a sip, then took a seat beside Eldon. The radio's jagged news transitioned to gentle music.

"So how goes the manuscript?" she asked. It was a query that seemed out of context. Eldon did not answer as the two sat in silence. Suddenly, Bev stood and said, "I feel like dancing."

Eldon smiled and got up. He took hold of her, and they swayed about the room.

"Have you thought where all this might be going?" she whispered. "You need special caring for."

"Well, so do you," Eldon said, looking into Bev's soft and liquid eyes. He leaned further ahead and kissed her forehead.

Bev's grasp tightened. "I have such up and down days...but with Erik now maybe being able to pick up his life...that'll help."

Eldon remained silent. When the song ended the two again sat on the davenport while Bev finished the liquor. She then got up, kissed him once more and moved towards the door. After one last smile, she was gone. Eldon then went to the kitchen table and sat down at the Royal, which had already been loaded with a sheet of paper. He increased the radio's volume to block out the neighbor's TV noise that was finding its way through the wall. On the radio a glib DJ was pushing his commercial wares:

> *"Remember folks, com' on by Freemont Ford Saturday. Bring the kids in for hotdogs and balloons. And remember, with the purchase of a new Ford Galaxy 500 receive twenty-five pounds of USDA approved beef. It's Ford-o-rama days. Don't miss it..."*

5

Days, then weeks passed. Eldon continued with an unwavering determination to complete the first phase of the writing project. But concentration had become a premium. Many nights dreams upended attempts for effective sleep, leaving the pillowcase wet and moist sheets crinkled. On this particular night during the last week of September, he was wide awake at two-fifteen in the morning. He climbed from bed and headed for the kitchen where he opened the refrigerator, hauled out a *Coca-Cola,* and took it to the table. Caffeine had been his drug of choice at the VA hospital, and the habit continued during the writing project. But, on this night Eldon craved a shot of dark rum to go with the soft drink, perhaps thinking auld lang syne might lend exactitude in getting him back square with the world. But he fought off the craving.

In Southeast Asia rum had been his fallback, a crutch when trying to muster a tolerable balance to the war's questionable purpose and his involvement in it. Rum also neutralized the paranoia effects coming from tokes on marijuana cigarettes. Drugs playing upon drugs had become a balancing medicinal. And now in reliving those scenes in his writing, and doing it sober, he figured his subconscious was simply raising hell, which in turn affected his sleep. His subconscious raising hell was something that had begun at West Port Veterans Medical Center. Still, Eldon persevered:

...Lt. Williams had led his men into a trap. It was a dangerous time for a second Louie. The shining young man from the Palose had enjoyed over a dozen search and destroy missions and seemed to have gained the confidence of his men—and

himself. But today the undergrowth danced with an uncertain breeze, a breeze doing little to moderate the stifling heat. I'd seen the setting before: the afternoon sun creating shadows; shadows giving way to dangerous imaginations the enemy had appeared from nowhere—then quickly vanishing within a bucolic setting populated by gentle peasants.

As our platoon took light but measured steps, Lt. Williams and radioman Jensen were on my right flank. I was on point, my M-16 locked and loaded, carried at a loose port arms. Something felt inevitable. It was an innate sense one sometime got after six months on search and destroy missions—perhaps something palm readers think they see in outstretched hands or very insecure, God-fearing old people might feel on the verge of death's imperative.

RAT-TAT-TAT-TAT, RAT-TAT-TAT TAT, RAT-TAT-TAT TAT.

The sudden incursion violated any preconceived notions of the kind of day we were to have. Instinctively, we hit the deck. The lieutenant looked over at me, motioning me to spread further out on his flank, an angled vector that might help me identify where shots might be if they rang out again. By now I knew my job well, but today I froze. It was the first time in six months my reflexes stalled.

"Get up on the angle goddamned it! Give us cover over here," Lt. Williams shouted. Jensen's head was down, barking coordinates in his radio set. The new private who had replaced Phillips was crouched just behind me, awaiting my movement so he could tail me. He had been in country just long enough to know his role.

When a very frustrated Lt. Williams stuck his head up once more to look sideways at me, a single round blew into his head. It was over fast. What followed was pandemonium as bullets flew crisscross in the rainforest, my platoon exacting 50-caliber vengeance. Expletives were hurled like fiery tracer *rounds in the spray of bullets. Jensen's barking into the radio handset intensified, as we now desperately needed an evac chopper. I was numb. I had let my lieutenant down. I had not done my job. And the lieutenant was dead...*

Eldon rolled the sheet of paper from the typewriter. The manuscript was ready, at least the initial phase of the project. He had given a self-imposed deadline for the story's segment to be completed before his afternoon appointment with counselor Doug.

6

Autumn's change of season pushed across the city, moderating temperatures. Change too had come to the Peter Pan. A topsy-turvy employee uprising had occurred as three strategic people quit—all on the same day: Walter; the head cook; and a morning-shift waitress. The trio had jumped ship, landing on their feet at other establishments around the city. Mrs. Butler's entrepreneurship was taking on water, and she had lost her love interest – those details murky. Rumor was Wes had been involved in a furious fight with an unknown assailant. In its aftermath, police found the unconscious Wes beat to a pulp and in possession of a small caliber pistol—a handgun he argued had been planted. With such an extensive rap sheet, any veracity of his argument quickly dissolved, and his parole was revoked. He'd be heading back to the *point-of- the-mountain.*

Walter was now at the Chesapeake, a nearby competing establishment and with it came a renaissance in attitude: "Like found money," he quipped. Walter happily toweled the bar and beer spigots, albeit a much smaller, less thirsty establishment than the Peter Pan. The Chesapeake's claim to fame was a staid Saturday afternoon coterie of grey-haired codgers who played Pinochle; it lent a well-behaved air to the place. The Chesapeake had fewer pool tables and pinball machines but a snappy kitchen trade. The place and become a new rendezvous for Eldon and Bev. Bev especially liked its intimacy. A cozy corner booth kept their conversation private as both continued to work on their relationship.

The timing of this change of venue also marked the completion of Eldon's initial writing project. As he entered the VA counselor's office with the enveloped manuscript tucked under an arm, Doug was taken aback.

"You been in a fight, Sergeant Burdett?"

"It's nothin'."

"With that face you surely couldn't have just walked into a door. We can talk about it if you like."

"No need."

"Sure?"

"Sure."

Doug opened the packet and thumbed through the pages. "I'm impressed by the quality of your composition after you gave me a few sample pages at our last meeting. Do you think this practice has helped?"

Eldon sat pensively, an index finger patrolling a crusted-over laceration on his chin. "Sure, like you've been saying all along, it's good to get things out in the open."

"How are the dreams?

"Every now and then one ambushes me."

"As I have told you, you may need to live with them, and that's why I want you to continue with your compositions. For some people, writing can be more helpful than group talk. You have skill as a writer, so I'd like you to keep it up, especially detailing the aftermath of your return to civilian life. This stuff is publishable, especially with the contentious war politics of the day."

Eldon shrugged his shoulders.

"How are other things?"

"She's fine. Don't know what she sees in me."

"We all have our qualities, Sergeant Burdett. And I can vouch for yours." Doug shaped a reassuring smile. "You're gonna get through this and from what you've told me of her, she's a big help."

"She has her demons."

"We all do. What's important is our awareness of them." There was a moment of awkward silence. "You really should see a doctor about that face."

"There's no broken bones. I had worse in the Nam. It'll heal... most things heal."

"What has your gal said about it?"

"She hasn't seen me in a week, but she's going to make me supper at her place tonight. We get together a few times a week. It's good we both have our space."

"Okay, now I want you to keep this writing up. Detail your experiences in West Port Veterans Medical Center. What's important is that you're ferreting out all the entrapments. Those tangles need to get straightened out and writing about them works for many. See you next week, but you have my card; it's my home number...call me anytime if you need to...anytime."

Eldon left the VA and caught a bus for town. Once in the city, for inexplicable reasons, he returned to the Peter Pan. There, he paused at the pool hall's top step, studying the blue Marlin trophy. The old fish's lone marble eye, dark brown and cloudy, appeared to have taken on an even darker shade. He took the steps down, pulled open the door, entered, and walked to his old bar stool where he ordered a *Coca Cola*. Neither patron nor employee knew one another's name. Two quarters were slid across the bar. A glance to the back of the place found Sammie and Brooks at a snooker table. Conspicuous laughter rose above a country/western song playing on the jukebox. Eldon turned back to his drink. He didn't feel like saying hello.

The usual cigarette smoke hung in the air, but the place was filled with a host of unrecognizable men. One poor soul in greasy railroader overalls sat two stools down, talking to himself. Beyond him was a new assembly of unkempt tavern philosophers who were testing their bibulous hypotheses. Their rants were porous in logic and a butchering of the English language. Eldon quickly drained his drink, feeling he had made his peace with the place. He left to catch the 10 Poplar Grove bus.

Home now, Eldon sat at the typewriter, sending out a burst of keystrokes. Finished, he removed the sheet of paper, placed it in a business-sized envelope, licked the fold, and penned it to Bev. He leaned the envelope against the typewriter. He then stood, walked to the closet and slipped into a Levi jacket, carefully feeling the pocket. Yes, it was where he had left it. After a long look around the place, he departed.

Eldon walked the short distance to Pioneer Park. In his face the sun's shimmering glare interplayed with afternoon shadows cast from thinning Elm trees. The sun's inevitable winter journey south was at hand. His route took him kitty-corner across the park, a park known for its robust stock of desperate, down-and-outers—an even more fraught gathering place than the Peter Pan. Eldon's racing mind conflicted with the slow and steady measure of his footsteps. Such bipolarity made perfect sense to a man trapped in a self-deprecating reality. After all, hadn't he declared himself to be nuts?

In the fashion of a bird of prey spotting a rodent, a scarecrow-clad man swooped in from nowhere, forcing a halt to Eldon's journey. The vagrant's face was swollen and grizzled, eyes bloodshot. He said he was working his way towards another bottle and asked for some change. Eldon fished for a coin and handed over a fifty-cent piece.

"God bless," was the beggar's reply. The man grasped the coin, turned and walked off.

"For everyone there is a season," Eldon called. There was no reply.

A chill was in the air, and Eldon pushed the collar of his Levi jacket about the neck. A honey wind was blowing, and it sent wayward Elm leaves skipping across the matted brown grass. Winds that blew in from the north country were a harbinger of a cold winter ahead, and thoughts of it carried to the Johnny Cash/Bob Dillon song, *North Country*, popular on the Peter Pan's jukebox.

The journey resumed, eventually ending at the 4th South Street viaduct where Eldon squatted amongst a labyrinth of vapid, concrete pillars. For a moment he felt secure, hidden, anonymous. Overhead, the hollow thuds of vehicular traffic rumbling caused vibrations to be felt deep in his chest. Slipping his hand into a

M. L. Graham

pocket, Eldon found a mini-bottle of bourbon. He hauled it out, twisting the cap to break the seal. The bottle was tipped. A hard swallow brought on a cough and a spluttering of the lips. It was the first drink he had had in over two years, and the liquor burned all the way down. Scents from its aftertaste reminded him of Bev.

Eldon thought of his lieutenant, the shiny young man from the Palouse. A quick learner, Lieutenant Williams had blended well with the platoon, a heroic mix of jungle smarts and dead reckoning skills in keeping germane the riddle of the quick and the dead. No question, the man had been the best platoon leader to date. Replays of the ultimate skirmish—and Eldon's inexcusable gig in it—played out once more. If only he had covered the officer's flank as he had been trained to do. No script rewrite could ever change that so why stew about it?

Bev's image pushed away those of the lieutenant's. Eldon had fought off a desire to love her. How could that work with a such a damaged man? She'd be fine now. She'd be fine. She'd soon have her son back, and that would make her happy. She deserved to be happy. She deserved so much more than what life had thrown her way. Bev's image grew even more focused as alcohol and a surge of adrenaline caused an uptick in Eldon's heart rate. Breathing came quick now. Again, the little bottle was tipped, its contents drained. The haunting Liz Damon melody rattled about in his head. *Loneliness Remembers What Happiness Forgets.* Bev loved the song.

The sudden cooing of a pigeon broke the tension. The bird moved closer to Eldon, the pigeon head-bobbing while performing a circular, ceremonial dance. Eldon studied it for a moment. "Hi, pretty birdy," he said in a soft whisper. "I've got no seeds…sorry." Eldon looked away. It was time.

Eldon reached into his jacket's pocket, fingering the cold steel of a handgun. But the act was interrupted by another shadow, this one overtaking him from behind. Startled, the pigeon took flight, its agitated flapping causing Eldon to reflex. He quickly pulled his hand from the pocket as the bird's departure set its down fluff adrift.

"Saw ya settin' from over there young fella and said to myself…
now there's a guy I taint never see'd here before."

Eldon eyed the man, his mind racing with a pounding heart.

"So, I sez, figurin' you'd just got off the U.P. what come past…
always slows to a stop here, ya know…needs to switch some before
it can move up the line. Lots of skids get off her here." The man
chortled, the phlegmy kind. He hawked and spit. "Sees 'at's how's I
come to the place. I camp here. See, right over there." He pointed. A
bed roll, tucked against a viaduct pillar, rested amongst a tangle of
trash. "That is until I gets run off by the cops what rummage about…
but sees I know when they pace the place, and I'm ready for 'em…"
Seemingly pleased with the pronouncement, he again chortled.

The rhythm of Eldon's intent had been torn asunder.

"You want a snort young fella? See'd you a minute ago tipping 'at
pee-wee bottle you had." This time his breathy chuckle came with
body movement as the man stooped on his haunches, apparently
meaning to stay a while. "I got a bigger bottle than that pee-wee deal."

Eldon said nothing as the old man fished for a half-pint whiskey
bottle from his tattered Eisenhower jacket—the olive-drab kind
sold at army/navy surplus stores. Still exposed loose threads on
the jacket's shoulders marked where rank and company patches
had once been.

"So, what's your story? Everybody's got one." The tramp took a
snort, reacting to it with a twist of his nose and a sideways pull of
the jaw. He wiped his mouth with the back of his hand and passed
the bottle to Eldon. Eldon refused it. "See mine goes all the back
to when I was probably your age. We was at war then. They tell me
we still at war…hell, war never ends does it, young fella. But 'hat
war going on now with 'em Chinamen what's in the newspapers,
ahh, that taint nothin' compared to what I saw."

By now Eldon's mind had decoupled from a manifestation of
pity and the need for self-annihilation to that of recapitulation.
"What the fuck!" he said under his breath, his cognitive acuity a
jumble of mush.

M. L. Graham

"See, today most of 'em youngin's ain't knowin' what the big war was…well let me tell you, I did. You bet I know. I was with the armored boys in the Libyan desert, '43 it was. Now that was war, son. And we know'd who we was fighting too. Em bugger Wehrmacht fought you toe-to-toe in 'at war…ain't like today when 'em Chinamen pop up like Jack-in-the boxes then disappear lickety-split."

Eldon's first utterance was, "I knew someone who was in that theater of war, old timer. My dad…and my lieutenant's dad in Vietnam. Both dads served with an armored division."

"Well, you don't say. They was in the armored?"

"Yeah. My dad talked a lot about it when I was growing up."

"He'd know…then he'd know." The old man tipped the bottle once more. "So, you sez you was in that Chinamen war?"

"I was."

"You lose men?"

"I did. One in particular."

"Lost some good men in my war too young fella." The man set the liquor bottle on the ground and found paper and tobacco in his pocket. He rolled a cigarette, a slight palsy to his hand. "Guess I don't look much like a war vet now. Life's funny…two sides of life's coin. That and booze can bring a good man to his knees. See for a long time I felt sorry for myself. I returned from the war in early '46 and settled down like a lot of us boys did…got married…had a family. But I never had no good luck with it…bad dreams…then picked up with the booze." He eyed the half-consumed liquor bottle with a philosophical tilt of the head as he fired up the cigarette. He took deep drag, held the smoke in his lungs, then expelled it through his nose and mouth. "It's a story repeated over and over again."

As if in a new dawn—an epiphany—Eldon's mind came into focus. He gained his feet, turned, and began walking away.

"Ya leaving young fella? Where you gonna go now?"

Eldon's turned back. "I have someone important to see." He pulled out his wallet, finding a twenty-dollar bill, and offering it to the man.

"I don't want your money young fella…or your sympathy. I was the cause of where I find myself today."

"Take it," Eldon said. He bent down and stuffed the bill into the man's jacket pocket. "We're war brothers now, old timer, a generation removed." Eldon then took out a card from his wallet. "Here. And go see this guy. You're a veteran. You're due benefits. This guy's name's Doug, and he helps guys like us." Eldon grew emotional, on the verge of crying. "And you helped me, old timer. This I owe you…more than you'll ever know."

The tramp studied the card for a moment then tucked it in the pocket with the folded cash. He took another deep draw on the cigarette.

"Now I have someone important to go see…someone very special. You take care. Best of luck to you. For everyone there is a season." Eldon flashed the victory sign.

The man reciprocated the gesture and said, "And a time for every matter under heaven."

ELDON'S PATH TOOK HIM IN THE DIRECTION OF THE PARK'S pavilion where he stopped at a trash barrel. After a furtive look around, he fished about the receptacle for a brown paper bag. He removed an empty liquor bottle from the sack then slipped the handgun from his pocket. He quickly removed the single bullet from the chamber and put it in his pocket. Eldon then took his handkerchief and wiped the pistol clean of fingerprints. That done, he slipped the weapon in the bag and stuffed the bag deep into the can. After another quick look around to confirm no one had seen the act, he left the park.

By the time Eldon reached Bev's apartment, late afternoon shadows had succumbed to the black of total darkness. He climbed the stairs to the second floor and knocked on the door. It opened.

"What happened? Your face!" Bev gasped.

"I've been to hell and back, Bev. But I met a man, an old man, a ghostly man of my future, and he helped me climb out of the pit I put myself in."

Bev thew herself at Eldon, grasping him tightly.

Eldon said, "You once told me I needed special caring...I think I'm ready for that."

Bev's eyes flooded with tears. "Here I am, Eldon, here I always am." She helped the wounded soldier inside, and the door closed quietly behind them.

Made in United States
Troutdale, OR
04/20/2024

19296815R00056